Springer Series in Reliability Engineering

For further volumes:
http://www.springer.com/series/6917

M. Luz Gámiz · K. B. Kulasekera ·
Nikolaos Limnios · Bo Henry Lindqvist

Applied Nonparametric Statistics in Reliability

Springer

Assoc. Prof. M. Luz Gámiz
Depto. Estadistica e Investigacion
  Operativa, Facultad Ciencias
Universidad Granada
Campus Univ. Fuentenueva
18071 Granada
Spain
e-mail: mgamiz@ugr.es

Prof. Nikolaos Limnios
Centre de Recherches de Royallieu,
Laboratoire de Mathématiques Appliquées
Université de Technologie de Compiègne
BP 20529
60205 Compiègne
France
e-mail: nlimnios@utc.fr

Prof. K. B. Kulasekera
Department of Mathematical Sciences
Clemson University
Clemson, SC 29634-1907
USA
e-mail: kk@clemson.edu

Prof. Bo Henry Lindqvist
Department of Mathematical Sciences
Norwegian University of Science and
  Technology
7491 Trondheim
Norway
e-mail: bo@math.ntnu.no

ISSN 1614-7839

ISBN 98-1-4471-2634-8   ISBN 978-0-85729-118-9 (eBook)

DOI 10.1007/978-0-85729-118-9

Springer London Dordrecht Heidelberg New York

British Library Cataloguing in Publication Data
A catalogue record for this book is available from the British Library

Cover design: eStudio Calamar, Berlin/Figueres

Printed on acid-free paper

Springer is part of Springer Science+Business Media (www.springer.com)

Preface

This book concerns the use of nonparametric statistical tools for the inferences of the performance characteristics of reliability dynamic systems operating in a certain physical environment that determines their behaviour through time.

Although many statistical methods rely on assumptions about the structure of the data to be analysed, there are many practical situations where these assumptions are not satisfied. In such cases, it may not be appropriate to use traditional parametric methods of analysis. In fact, very often a free-model method, and therefore, a data driven focus to the problem, is the only option.

The term nonparametric does not mean that there is a complete lack of parameters; rather it implies that the number and the nature of the parameters are not fixed in advance. In any case, nonparametric methods require a very limited number of assumptions to be made about the underlying distribution of the data, which undoubtedly confers an advantage over parametric methods. First of all, they have a wider applicability and are more robust. Moreover, nonparametric methods are intuitive, flexible and simple to carry out. Among the disadvantages, we should mention that appropriate computer software for nonparametric statistics may be limited; however, this is a continually improving situation.

Roughly speaking, we could summarize the main purpose of this book as understanding systems' performance. To undertake this task, there are many aspects that manufacturers, analysts and of course, users must take into account such as: the mode and the intensity of usage, the environmental conditions, the maintenance actions (when these are accomplished), the internal structure of the system, etc. To what extent all these features should be examined is mainly a question of the quality and quantity of the information the analyst is willing to get. Because of this, we may be confronted with many different practical situations that vary between two extremes.[1] On the one hand, the less informative procedure applies when the method used for system observation does not make use of any

---

[1] Nicolai, R.P., *Maintenance Models for Systems Subject to Measurable Deterioration*, Ph.D. Thesis, Erasmus University, Rotterdam (2008)

particular knowledge of the physical characteristics implicit in the system operation. In this case, the deterioration model consists merely of describing the relation between time and failure, and is therefore reduced to a lifetime distribution model. We refer to these models as black-box models and we devote the first part of this book to study these models by using nonparametric approaches.

In contrast, we may be facing situations in which a deeper (as far as possible) understanding of the different mechanisms of failures is required and therefore, we need models that are as close as possible to a full description of the real system, in other words, a physics-based model is required. Think for example of the case where the different components of the system, as well as the relationships between them, are relevant to get significant information for understanding the system behaviour and consequently for making certain decisions in the manufacture procedure. These models are called white-box models in the literature. Although the major advantage of these models is their enormous degree of flexibility, nevertheless, the extremely high complexity and computing burden involved in setting a pure white-box model as a copy of reality make their formulation not feasible at all in practice. The third part of the book discusses how to represent the evolution to failure of reliability systems under three approaches that could be considered, to some extent, into this group.

The intermediate formulation is what we denote as grey-box models. We understand that a grey-box model provides a mathematical representation, but some of the physics is approximated. For our purposes, the model is reduced to a deterioration model that indicates the time-dependent deterioration and failure. These models are studied in the second part of the book.

This book is primarily intended for practitioners and researchers in reliability engineering, who faced with reliability data would like to explore the possibility of nonparametric descriptions of underlying probability mechanisms. The book is hence an alternative to much of the current reliability literature which is based on parametric modelling. Given the extensive collection of models and methods considered, the book should also be of interest for a wider audience working in for example software engineering, biostatistics, operations research, economics and other fields where events occurring in time are studied.

Computations and graphs are done using R (http://www.r-project.org/) and MATLAB. Some of the program code and datasets can be obtained from the authors.

We are grateful to all our collaborators and colleagues for their contributions to the research and for stimulating discussions and advice.

Granada, July 2010   M. Luz Gámiz
Clemson, July 2010   K. B. Kulasekera
Compiègne, July 2010   Nikolaos Limnios
Trondheim, July 2010   Bo Henry Lindqvist

Contents

Acronyms

AMISE	Asymptotic Mean Integrated Squared Error
ARP	Alternating Renewal Process
CDF	Cumulative Distribution Function
CV	Cross-Validation
EMC	Embedded Markov Chain
HPP	Homogeneous Poisson Process
i.i.d.	Independent Identically Distributed
ISE	Integrated Squared Error
KM	Kaplan-Meier
MISE	Mean Integrated Squared Error
MRE	Markov Renewal Equation
MRP	Markov Renewal Process
MSE	Mean Squared Error
MTTF	Mean Time to Failure
MTTR	Mean Time to Repair
MUT	Mean Up Time
NHPP	Non-Homogeneous Poisson Process
ORP	Ordinary Renewal Process
PAVA	Pool Adjacent Violators Algorithm
PL	Partial Likelihood
pdf	Probability Density Function
RCM	Right Censoring Model
rocof	Rate of Occurrence of Failures
RP	Renewal Process
r.v.	Random Variable
SMP	Semi-Markov Process
TRP	Trend-Renewal Process
$\xrightarrow{d}$	Convergence in distribution
$\xrightarrow{a.s}$	Almost sure convergence
$*$	Convolution operator
$\lvert \cdot \rvert$	Absolute value function

$[\cdot]$	Integer part of a real number
$\prec$	The usual partial ordering defined in $S=[0,1]^r; x_1 \prec x_2 \Leftrightarrow x_{1i} \le x_{2i}, i = 1,2, ..., r$
$\mathbf{M}^{\mathrm{T}}$	Transpose of a matrix M
$\mathbf{0}$	Vector of 0's that denotes the *complete failure* state of a reliability system
$\mathbf{1}$	Vector of 1's that denotes the *perfect functioning* state of a reliability system
A	Steady-state or limiting availability
$A(t)$	Instantaneous availability
A_{av}	Average availability
B	Bias of an estimator
b	Bandwidth parameter
C	Censoring random time
$D(t)$	At-risk process; $D(t) = I[\ge t]$ for any $t>0$
$\frac{d^m}{dt^m}$	Derivative function of order m, for $m \ge 1$
$\mathscr{E}$	Effective age function
F	Cumulative distribution function of a *r.v.* T; $F(t) = \Pr\{T \le t\}$
f	Density function of a *r.v.* T; $f(t) = \frac{d}{dt}F(t)$
$\bar{F}$	Survival function of a random variable T; $\bar{F}(t) = 1 - F(t)$
$\bar{F}_s$	Conditional survival function; $\bar{F}_s(t) = \bar{F}(s+t)/\bar{F}(s)$
$F^{(k)}$	k-fold Stieltjes-convolution function of F; $F^{(k)}(t) = F^{(k-1)} * F(t) = \int_0^t F^{(k-1)}(t-u)dF(u)$, for k>1
h	Bandwidth parameter
$I[\cdot]$	Indicator function
$1_{\{\cdot\}}$	Indicator function
K	Cumulative kernel function
k	Kernel density function
L	Likelihood function
l	Log-likelihood function
$\ln$	Natural logarithm
M	Renewal function of a renewal process or, Mean function of a counting process; $M(t) = E[N(t)]$
m	Renewal density of a renewal process or rocof of a counting process; $m(t) = \frac{d}{dt}M(t)$
$\mathbf{N}$	The set of nonnegative integers
$N(t)$	Cumulative number of events (renewals or failures) arriving in the interval (0,t].$\{N(t); t \ge 0\}$ is a counting process
n	Sample size
$\mathbf{P}_i(\cdot)$	Semi-Markov transition function
Q	Semi-Markov kernel of a SMP ZZ $= (Z_t)_{t \in R_+}$
$R(t)$	Reliability function
$\mathbf{R}^+$	The set of nonnegative real numbers
r	Number of components of a relibility system
$ro(\bullet)$	Rocof (Chap. 6)
S_k	Waiting time until the kth renewal cycle is completed or the kth failure has occurred

s	Spline function
T	Time to failure or lifetime
T_i	Sample element ($i = 1, 2, \ldots, n$)
T_i^*	Bootstrap sample element ($i = 1, 2, \ldots, n$)
$T_i^\bullet$	Bootstrap sample element ($i = 1, 2, \ldots, n$)
$\bar{t}$	Reported sample mean; a realization of $\bar{T}$
$\mathbf{U}$	Random state-vector, that is $\mathbf{U} = (U_1, U_2, \ldots U_r)$
U_i	Random state of component labelled i, $0 \leq U_i \leq 1$
V	Random state of a reliability system
Var	The variance of an estimator
W	Cumulative kernel function
w	Kernel density function
$\mathbf{X}$	p-dimensional vector of covariates
X_j	The jth component of the covariate vector, for $j = 1, 2, \ldots, p$
Y	Minimum of the lifetime and the censoring time: $Y = \min\{T, C\}$
Z	Cumulative hazard function of an r.v. T, $Z(t) = -\log \bar{F}(t)$
Z	Semi-Markov process, i.e. $Z = (Z_t)_{t \in R_+}$ (Chap. 6)
z	Hazard function of an r.v. T, $z(t) = \frac{d}{dt}(-\log \bar{F}(t))$
$\boldsymbol{\beta}$	Unknown regression parameter vector
β_j	The jth component of the regression vector, for $j = 1, 2, \ldots, p$
$\widehat{\boldsymbol{\beta}}$	An estimator of $\boldsymbol{\beta}$
δ	The censoring indicator; $\delta = I[T \leq C]$
ϵ	Random perturbation
Γ	The Gamma function
γ	Conditional intensity function of a counting process
Λ	Cumulative hazard function of an r.v. T, $\Lambda(t) = -\log \bar{F}(t)$; cumulative trend function of TRP
λ	Hazard function of an r.v. T, $\lambda(t) = \frac{d}{dt}(\Lambda(t))$; trend function of TRP
μ	Unconditional intensity function of a counting process
μ_T	Expected value of an r.v. T
∇	The gradient operator
ω_i	Weighting function
Φ	Cumulative distribution of the standardized Normal law
ϕ	Density function of the standardized Normal law
ψ	Markov renewal function
σ_T^2	Variance of an r.v. T
$\varphi(\cdot)$	The structure function of a reliability system
$\hat{\varphi}(\cdot)$	Smooth estimate (not necessarily monotone) of $\varphi(\cdot)$
$\tilde{\varphi}(\cdot)$	Monotone smooth estimate of $\varphi(\cdot)$
ζ	Critical state of the system
θ	Unknown parameter
$\hat{\theta}$	Estimator of θ
$\hat{\theta}^*$	Bootstrap version of the estimator $\hat{\theta}$
$\hat{\theta}^\bullet$	Bootstrap version of the estimator $\hat{\theta}$

Part I
Black-Box Approach:
Time-to-Failure Analysis

Our objective at this stage is to model the relation between the time and the failure when no information regarding the operating conditions are not incorporated into the data collection. Our approach is to study the random time til failure of an item or a system. The statistical techniques at this level are primarily focused on making inferences on the failure (hazard) rate function, cumulative failure rate function and the reliability (survival) function of the random variable that measures the time to failure. We identify numerous censoring/truncation schemes that arise due to natural data collection constraints and modify the analyses to accommodate such incompletenesses.

We start with the classical Kaplan–Meier estimator for the survival function and the Nelson–Aalen estimator for the cumulative hazard function for a lifetime random variable. Then we extend the discussion to smoothed versions of these estimators which are more in line with the very frequently made assumption that the lifetime variable possesses a probability density function. This discussion is then followed by an introduction to smoothing techniques such as kernel smoothing, local linear method, spline method, etc. for the estimation of failure rate functions in the presence of censoring. An inherent issue in all smoothing methods is the choice of the smoothing parameter (bandwidth). This aspect is visited with some detail where we describe a few popular and theoretically sound bandwidth selection techniques for estimating the underlying density function and the failure rate function.

Part I
Black-Box Approach:
Time-to-Failure Analysis

Chapter 1
Lifetime Data

1.1 Smoothing Methods for Lifetime Data

1.1.1 Introduction

Engineering and medical studies often involve measuring lifelengths of individuals or various items. These values are modeled as non-negative random variables. The analysis of various characteristics of the underlying probability distributions of such random variables is often required in practical decision making. For example, in comparing several treatment protocols for patients with cancer, the average remission time for each treatment can serve as a useful tool in deciding the most appropriate treatment.

To make things formal, suppose T denotes a lifelength. In this work, we define the cumulative distribution function F of T by $F(t) = P[T \leq t]$ and assume that the probability density function f (the derivative of F) exists.

In general, F (and therefore f) is unknown. In many applications, it is of interest to estimate certain quantities that are solely determined by the cumulative distribution function F. These are typically referred to as functionals of F, which we denote by $L(F)$. These functionals are very valuable decision-making tools in most lifetime studies. A simple yet commonly used functional is the mean of T, which is defined as $\mu = L(F) = \int t \, dF(t)$. In some instances, one may be interested in estimating the tail probability

$$P[T \geq t_0] = \int_{t_0}^{\infty} dF(t)$$

for a given t_0, where T is the lifelength of an item following the cdf F. In the lifetime data literature, this tail probability is called the reliability (survival

M. L. Gámiz et al., *Applied Nonparametric Statistics in Reliability*,
Springer Series in Reliability Engineering, DOI: 10.1007/978-0-85729-118-9_1,
© Springer-Verlag London Limited 2011

probability) at time t_0. Alternatively, it may be of interest to evaluate the hazard rate (failure rate)

$$\lambda(t) = \lim_{x \to 0} \frac{P[t < T \le t + x | T > t]}{x}$$
$$= f(t)/[1 - F(t)]$$

corresponding to the cdf F. Here, we assume that the density function f is well defined for all $t \ge 0$. The hazard rate gives information on instantaneous failure probability of an item of age t. This index is very useful in understanding the instantaneous behavior of a functioning system. For example, a system that has a λ that is monotone increasing is considered to have positive aging (older the item, weaker it is in some sense), whereas items or populations that exhibit negative aging (strengthening with age) are modeled with failure rate functions that are decreasing in time.

Among other commonly used reliability indices that are of the form $L(F)$ for a lifetime distribution F are the mean residual life function and the percentile residual life function. Lawless [16] provides a detailed discussion on various reliability indices and their relationship with concepts of aging.

To estimate a functional $L(F)$, a common approach is to substitute a reasonable estimator of F in place of F in $L(F)$. If the functional form of F is known up to a set of parameters, i.e., $F(t) = F(t, \theta)$, one may use a standard parametric estimation method to estimate the unknown parameter vector θ by $\hat{\theta}$ and use $L(F(., \hat{\theta}))$ as an estimator of $L(F)$. Popular and effective parametric methods include the likelihood method, method of moments and various minimum distance methods such as M-estimation. In cases where the functional form of F is unknown, the case we are examining here in detail, one resorts to nonparametric techniques to find an estimator of F. Once a suitable estimator F_n (say) of F is obtained, we can estimate $L(F)$ by $L(F_n)$. In the following, we discuss several nonparametric estimators based on different types of data for various reliability indices that are of the type $L(F)$.

1.2 Complete Samples

In a lifetime study, experimenters place item or individuals on test and observe the failure (event) times on these individuals or items. For example, if we are interested in studying the properties of the probability distribution of remission times of patients with cancer, we observe the data for several individuals with homogeneous characteristics. If we have to further categorize the behavior of lifetime distributions due to non homogeneity among the patients, we resort to regression models that can accommodate such differences. This aspect is discussed in Chap. 7. In this section, we examine the case where all the observations are available to the experimenter. In reliability and survival analysis literature, this is referred to as the complete sample case.

In general, the data $T_i, i = 1, \ldots, n$ in a study of the above type can be thought of as a simple random sample from a life distribution F; i.e., the observations are independent and identically distributed with a common cumulative distribution function F. The simplest, yet efficient, nonparametric estimator of F is the empirical distribution function F_n defined as

$$F_n(t) = \frac{\sum_{i=1}^{n} I[T_i \leq t]}{n}, \tag{1.1}$$

where $I[A]$ is the indicator of the event A. It is easy to show that

$$E[F_n(t)] = F(t)$$

and

$$\text{Var}[F_n(t)] = \frac{F(t)(1 - F(t))}{n}.$$

Hence, if one were interested in estimating a functional $L(F)$, a straightforward nonparametric estimator is $L(F_n)$. For example, if $L(F)$ is the population mean μ_T for the random variable T, then $L(F_n)$ is the sample mean $\overline{T} = \sum_{i=1}^{n} T_i / n$.

The estimation of the density function f based on a random sample is most commonly done using a kernel density estimator. A kernel density estimator can be motivated as follows. First, consider approximating the derivative f of F at a given t. This approximation can be written as

$$f(t) \approx \frac{[F(t + h) - F(t - h)]}{2h}$$

for a small increment h. Now, since F is estimated by F_n, it is reasonable to estimate f by

$$f_n(t) = \frac{F_n(t + h) - F_n(t - h)}{2h}.$$

This f_n can be rewritten as

$$f_n(t) = \sum_{i=1}^{n} \frac{I[-1 < \frac{T_i - t}{h} \leq 1]}{2nh}.$$

Thus, the estimator f_n is written using a uniform density on $[-1, 1]$. Parzen [22] generalized the above estimator using any probability density function k instead of the uniform density function on $[-1, 1]$. In particular, a kernel density estimator is defined as

$$f_n(t) = \frac{1}{nh} \sum_{i=1}^{n} k\left(\frac{T_i - t}{h}\right). \tag{1.2}$$

Here, k is called the kernel function and h is called the smoothing parameter (the bandwidth). Note that the above estimator can actually be written as

$$f_n(t) = \int \frac{1}{h} k\left(\frac{u-t}{h}\right) dF_n(u) \qquad (1.3)$$

where F_n is the empirical cdf defined above. This estimator possesses desirable large sample properties when the underlying true density function f is smooth with a continuous second derivative. The most common assumptions on the kernel function k for consistent estimation of a density f are

A1: $k \geq 0$ is a symmetric probability function.
A2: $\int u^2 k(u) du = \kappa_2 < \infty$.
A3: $\int k^2(u) du = \eta_2 < \infty$.

Under the aforementioned smoothness assumptions on the density and assumptions A1–A3, we can derive the bias, the variance and the asymptotic distribution of the density estimator [24] as

$$E[f_n(t)] = f(t) + \frac{1}{2} h^2 \kappa_2 f''(t) + o(h^2)$$

$$\text{Var}(f_n(t)) = \frac{f(t)}{nh} \eta_2 + o((nh)^{-1})$$

and

$$\sqrt{nh}[f_n(t) - f(t)] \xrightarrow{D} N(0, f(t)\eta_2)$$

as $n \to \infty$. Based on the above expressions on the bias and the variance of f_n, the mean squared error of estimating the density function at a given time point t is given by

$$E[f_n(t) - f(t)]^2 = \frac{1}{4} h^4 \kappa_2^2 (f''(t))^2 + \frac{1}{(nh)} \eta_2 + o(h^4) + o((nh)^{-1}).$$

By a close examination of the leading two terms of the above mean squared error, we observe that a bandwidth of order $n^{-1/5}$ gives an optimal rate of $O(n^{-4/5})$ for the mean squared error to converge to 0. Kernel functions that satisfy the above assumptions A1–A3 are called second-order kernel functions. Further improvements in the rate at which the mean squared error of a density estimator converges to zero are discussed by Singh [28] by using kernel functions that may take on negative values (higher-order kernel functions). In fact, by a technique referred to as the generalized jackknifing (GJ), kernel functions can be constructed such that the squared bias term reduces in its order so that the convergence rate of the mean squared error can be made arbitrarily close to n^{-1} in a nonparametric density estimator. Schucany and Sommers [29] and Muller and Gasser (1979) provide details of GJ.

The empirical distribution function is a step function for all sample sizes. In estimating a continuous distribution function, it may be desirable to use a continuous estimator. A reasonable continuous estimator of the cdf F can be constructed using a density estimator. In particular, based on f_n above, we can construct an estimator of F as

$$\tilde{F}_n(t) = \int_{-\infty}^{t} f_n(u)\mathrm{d}u$$

$$= \frac{1}{n}\sum_{i=1}^{n} K\left(\frac{t - T_i}{h}\right) \tag{1.4}$$

$$= \int K\left(\frac{t - u}{h}\right)\mathrm{d}F_n(u) \tag{1.5}$$

where

$$K(x) = \int_{-\infty}^{x} k(u)\mathrm{d}u$$

is the cdf corresponding to the kernel function k. Reiss [26] has studied the properties of the convolution smoothed empirical distribution function $\tilde{F}_n$ above showing that the empirical cdf is "deficient" compared with the smoothed version above. In particular, we can compare the MSE of the smooth estimator $\tilde{F}_n(t)$ with the MSE (variance) of the empirical distribution function under some smoothness assumptions on F. Suppose that for a fixed real number t and $m \geq 1$, F satisfies

A4: $\quad \left|F(x) - \sum_{i=0}^{m} \frac{a_i}{i!}(x - t)^i\right| \leq \frac{A}{(m + 1)!}|x - t|^{m+1}$

for all x where $a_i, i = 0, \ldots, m$ are real numbers and A is a positive constant. This assumption allows F to be sufficiently smooth with continuous derivatives of mth order. Then, letting

$$M(t) = nE\left[\tilde{F}_n(t) - F(t)\right]^2 - F(t)[1 - F(t)],$$

we can obtain the inequality

$$\left|M(t) + a_1 h \int tb(t)\mathrm{d}t\right| \leq n\left[h^{m+1}\frac{A}{(m + 1)!}\int |k(t)t^{m+1}|\mathrm{d}t\right]^2 + O(h^2) \tag{1.6}$$

where $b(x) = 2k(x)K(x)$. Using the symmetry of the kernel k, we can then simplify the above inequality to show that

$$nE\left[\tilde{F}_n(t) - F(t)\right]^2 \leq F(t)[1 - F(t)] - n^{\frac{-1}{2m+1}}[\psi_m + o(1)] \qquad (1.7)$$

where ψ_m is a positive constant that depends on the kernel function k and A above. Hence, $\text{nMSE}[\tilde{F}_n(t)]$ can be less than $F(t)[1 - F(t)]$ showing the deficiency of the empirical cdf compared with the smoothed estimator of the distribution function F for certain choices of the bandwidth. The deficiency result above essentially states that the MSE of the smoothed estimator agrees with that of the empirical cdf up to a term of order n^{-1}, and it has a negative term of a lower order (i.e., $o(n^{-1})$), which can become significant for small sample sizes. Several simulation studies have shown this phenomenon for many distributions. However, it should be noted that the estimator $\tilde{F}_n(t)$ is biased for F although the bias vanishes fast as the sample size increases.

1.3 Censored Samples

Many lifetime experiments produce incomplete data. The most common incompleteness occurs due to censoring; some lifetimes are not observed, but either a lower bound (right censoring) or an upper bound (left censoring) on some lifetime measurements is available. For example, in right censoring, instead of observing $T_i, i = 1, \ldots, n$, the experimenter only observes $Y_i = \min\{T_i, L_i\}$ and $\delta_i = I[T_i \leq L_i]$ for a set of values $L_i, i = 1, \ldots, n$. Here, L_is can be either fixed or random and are called censoring times. For example, patients with cancer entering a study may leave at random points due to various reasons, and the complete set of remission times may not be available for these patients producing right-censored observations. Another setting in which one may observe right-censored data is when there are competing risks; the experimenter is interested in failures due to one reason, but the system may fail due to another cause.

Similar to right censoring, left-censored data arise in many practical settings. An example is the following. Suppose we were interested in modeling the age at which a child develops a certain skill. If we take a random sample of children from a given population, there is a possibility that some children may already have developed the skill that is being investigated. Hence, we are unable to observe the exact age at which the particular skill is developed for such children. Instead, we collect left-censored data where the observations are of the type $\max\{T, \tilde{L}\}$ and $\tilde{\delta} = I[T \geq \tilde{L}]$ where $\tilde{L}$ is the left censoring time.

We address the right censoring in detail here. However, the inferential methods discussed here can be easily applied to left-censored data models with obvious modifications. We shall assume that both the lifetimes and censoring times are independently and identically distributed. Let the cumulative distribution function for random right censoring times L_i be G, and suppose the cumulative distribution function of a Y is H. In our discussion below, we assume that the random censoring variable L is independent of the lifetime variable T.

When the data are right censored, the nonparametric estimation of F and f becomes somewhat difficult. For Y_i and δ_i, $i = 1,\ldots,n$ above, let $Y_{(i)}$s are the ordered Ys with the corresponding δ_is denoted by Δ_i. The estimation of the survival function S (hence the cdf F and associated functionals) in the presence of censoring has been widely discussed in the literature. Among one of the very early developments is the Kaplan–Meier [11] estimator, hereafter KM estimator, of $S = 1 - F$. For technical simplicity in almost all problems involving the survival function, density function and the failure rate function, we assume that the estimation is done in the interior of an interval $[0, \tau]$ where $\tau < \inf\{t : H(t) = 1\}$. A self-consistent version (Efron, 1967) of the KM estimator is given by

$$\widehat{S}_n(t) = \begin{cases} \prod_{j:Y_{(j)} \leq t} \left(\frac{n-j}{n-j+1}\right)^{\Delta_j} & \text{if } t < Y_{(n)} \\ 0 & \text{if } t \geq Y_{(n)} \end{cases} \tag{1.8}$$

This estimator is in fact the nonparametric likelihood estimator of the survival function S [16]. Unlike the empirical survival function $1 - F_n$, the KM estimator is biased. However, its bias decreases at an exponential rate as the sample size increases. Felming and Harrington [6] show that

$$E[\widehat{S}_n(t)] = S(t) + b_n(t)$$

where $0 \leq b_n(t) \leq F(t)[1 - H(t)]^n$ where the cdf H is defined above. Since we assume the cdfs F and G are both continuous, ties in lifetimes/censoring times have zero probability. However, in practice, there can be ties. If there were ties, the KM estimator can be modified as follows. Let the distinct failure times are $0 < t_1 < \cdots < t_k$ with n_i individuals at risk just before t_i and d_i failures at each t_i. Then, the KM estimator takes the form

$$\widehat{S}_n(t) = \prod_{j:t_j \leq t} \left(\frac{n_j - d_j}{n_j}\right).$$

The variance of the KM estimator is estimated using the classical Greenwood formula. For $t_1 < \cdots < t_k$ above, the variance of $\widehat{S}_n(t)$ is estimated by

$$\widehat{\mathrm{Var}}(\widehat{S}_n(t)) = \widehat{S}_n(t)^2 \sum_{t_i < t} \frac{d_i}{n_i(n_i - d_i)}.$$

Furthermore, $\sqrt{n}(\widehat{S}_n(t) - S(t)$ has an approximately normal distribution for large sample sizes. In particular,

$$\sqrt{n}(\widehat{S}_n(t) - S(t) \xrightarrow{\mathcal{D}} N(0, \sigma^2(t))$$

where

$$\sigma^2(t) = S^2(t) \int_0^t \frac{1}{H(s)} d\Lambda(s)$$

where $\Lambda(t) = \int_0^t \lambda(u)du$ is the cumulative hazard function. An alternative estimator of the survival function has been obtained via an estimated cumulative hazard function. Motivated by the fact that $\Lambda(t)$ is the accumulation of λ over the interval $[0, t]$, and using the definition of the failure rate, Nelson–Aalen estimator (ref) of Λ is given by

$$\hat{\Lambda}(t) = \sum_{i=1}^{n} \frac{\Delta_i I[Y_{(i)} \leq t]}{n - i + 1}. \qquad (1.9)$$

Now, using the relationship $\Lambda(t) = -\ln(S(t))$, it is natural to define an estimator of the survival function [3] as

$$\widehat{S}(t) = e^{-\hat{\Lambda}(t)}.$$

Both estimators $\widehat{S}_n$ and $\widehat{S}$ of the survival function S defined above are step functions. In the sequel, we shall introduce two versions of smooth estimators of the survival function and the distribution function.

1.4 Density Estimation with Censored Data

Given an estimator of the survival function (hence the cdf F), it has become a common practice to estimate the underlying density function f by straightforward generalizations of the kernel estimators given in (1.3). For example, for right-censored observations above, we can replace the empirical cdf F_n by $\hat{F}_n = 1 - \hat{S}_n$ and write an estimator $\hat{f}_n$ of f as

$$\hat{f}_n(t) = \int \frac{1}{h} k\left(\frac{u-t}{h}\right) d\hat{F}_n(u). \qquad (1.10)$$

Note that the estimator $\hat{F}_n$ is a step function that jumps at the ordered uncensored observations (i.e., for $Y_{(i)}$s with $\Delta_i = 1$). Letting the jump size at each $Y_{(i)}$ be s_i, we can rewrite the above estimator as

$$\hat{f}_n(t) = \sum_{i=1}^{n} \frac{s_i \Delta_i}{h} k\left(\frac{Y_{(i)} - t}{h}\right). \qquad (1.11)$$

As in the complete sample case, the bias, the variance and hence the MSE of the density estimator have been discussed by many authors. In fact, the consistency of this estimator $\widehat{f}_n(t)$ has been examined under very general conditions by many authors. Karunamuni and Song (1991) obtain the rate

$$\sup_{[0,\tau]} |\widehat{f}_n(t) - f(t)| = O_p(n^{-\frac{1}{3}})$$

for a second-order kernel function. In fact, as shown in Karunamuni and Song (1991), when one uses higher-order kernels, this rate can be reduced arbitrarily close to $n^{-1/2}$.

Since observed data in a lifetime experiment (lifetimes and the censoring times) are non-negative, it is desirable to have $\widehat{f}_n(t) = 0$ for $t < 0$. However, the above kernel density estimator can take positive values for $t < 0$ when a symmetric kernel is used. This is called a spillover problem for the density estimator. There are adjustments that can correct such effects. We will discuss the spillover problem and remedies in Sect. 1.6 of this chapter.

We can define a smoothed estimator of the distribution function using an integrated density estimator as in (1.4) above. One way to define a smoothed estimator of F is to define an estimator as

$$\widetilde{F}_n^*(t) = \int_{-\infty}^{t} \widehat{f}_n(x)\mathrm{d}x.$$

This estimator can be rewritten as

$$\widetilde{F}_n^*(t) = \int K\left(\frac{t-y}{h}\right)\mathrm{d}\widehat{F}_n(y) \tag{1.12}$$

Under the same assumptions that lead to the mean squared error inequality (1.7) above, it has been shown that the above estimator satisfies a mean squared error inequality that compares the MSE of $\widetilde{F}_n^*(t)$ with that of $\widehat{F}_n(t)$ [8]. In particular, for suitably chosen bandwidth sequences, the estimator $F_n = 1 - \widehat{S}_n$ is deficient with respect to the smooth estimator $\widetilde{F}_n^*$ in the sense of normalized mean squared errors as discussed in Sect. 1.1.1.

Motivated by the smoothing of the empirical distribution function [13], developed a smoothed estimator of the survival function as

$$\widetilde{S}_n(t) = \int \widehat{S}_n(u)\frac{1}{h}k\left(\frac{t-u}{h}\right)\mathrm{d}u.$$

This estimator also has a bias that decreases at an exponential rate as the sample size increases. In addition [13], show that the KM estimator $\widehat{S}_n$ can be deficient with respect to this smoothed estimator as the sample size increases for a suitably chosen smoothing parameter h.

One major drawback in the smoothed estimators of this type is the boundary bias near the origin. Note that all survival distributions are supported over $[0, \infty)$, and therefore when estimating the survival function (and other smooth functions) using symmetric kernels near the origin, the estimator can ill behave. This is due to the lack of data below 0. Note that the kernel estimator uses the data in a window with a width $2h$. Thus, for $0 < t < h$, the amount of observations is typically not balanced on both sides of the window around t giving rise to a bias, which is referred to as boundary effect (or boundary bias). This is not the same as the spillover effect that will be discussed in a subsequent section. In most situations, use of a slightly modified kernel function that depends on both t and h can eliminate this problem. Such kernels are called boundary kernels, and a detailed treatment of this issue can be found in Eubank [5].

We illustrate the smooth estimation of the survival function using the failure and censoring times of diesel engine fans [21]. These data give the lifelengths (or the censoring times) of 70 diesel engine fans measured in hours. The observations ranged from a minimum value of 450 h (a lifetime) to 11,500 h (a censored value) where 58 observations were censored. The plots of the smoothed KM estimator for several bandwidths are given in Figs. 1.1, 1.2 below. All three estimators behave very similarly for almost all t values except near the origin. The boundary effect near the origin is very clear for the two estimates using the bandwidths 0.025 and 0.035. Notably, the curve corresponding to the smallest bandwidth has the smallest bias near the origin.

Fig. 1.1 Survival probability $S(t)$ against time t (in 10,000 h units) for three bandwidths

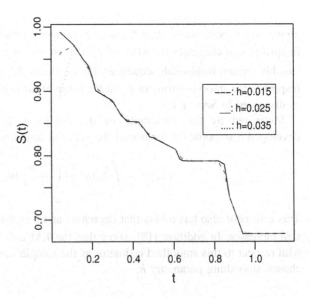

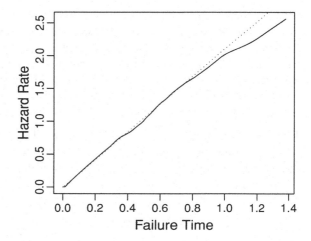

Fig. 1.2 Plots of failure rate estimators for simulated data. Hard line is the kernel estimator and the dotted line is the cubic spline estimator

1.5 Failure Rate Estimation

Given a density estimator and a survival function estimator, one can define an estimator of the failure rate. Using the KM estimator and the density estimator (1.11) above, a Watson–Leadbetter (1964) type estimator of the failure rate is defined as

$$\hat{\lambda}(t) = \frac{\hat{f}_n(t)}{\hat{S}_n(t)}, \quad t > 0. \tag{1.13}$$

Although the common choice in the denominator of (1.13) has been the Kaplan–Meier estimator $\hat{S}_n(t)$, it can be replaced by any reasonable estimator of the survival function. These ratio-type estimators are consistent for the underlying hazard rate λ over a large portion of the support of f.

Another ratio-type estimator for the hazard function is motivated using the following argument. Consider $W(y) = P[Y \leq y, \delta = 1]$. Then, it is easy to see that

$$dW(y) = [1 - G(y)]dF(y).$$

Now, consider $dW(y)/(1 - H(y)) = \lambda(y)$. Hence, if we can estimate $dW(y)$ and $1 - H$ separately, this will lead to a ratio-type estimator of λ. In this approach, the function H can be easily estimated by the empirical cdf H_n of the Ys. A kernel estimator of $dW(y)$ can be defined as

$$\widehat{dW}(y) = \frac{1}{nh} \sum_{i=1}^{n} \delta_i k\left(\frac{y - Y_i}{h}\right),$$

and therefore, an estimator of λ can be defined as

$$\tilde{\lambda}(t) = \frac{\widehat{dW}(t)}{1 - H_n(t)}. \tag{1.14}$$

   This estimator and the ratio estimator defined in (1.13) both have mean squared errors of the same order. However, the exact expressions for the MSEs are slightly different. The ratio-type hazard rate estimators heavily depend on the density estimator and the survival function estimator used in the ratio. This type of a ratio estimator has also been suggested by Marron and Padgett [18] where they used a density estimator based on the relationship between S, $G^* = 1 - G$ and $1 - H$ used in constructing the estimator (1.14) above. Note that $dW(t) = f(t)G^*(t)$, and therefore, using the same estimator $\widehat{dW(t)}$ to estimate $dW(t)$, one can estimate f by $\widehat{dW(t)}/G_n^*(t)$ for an estimator G_n^* of G^*. An estimator of G^* can be obtained by a KM estimator by reversing the roles of the observations and the censoring values. In particular, we can define

$$G_n^*(t) = \begin{cases} \prod_{j:Y_{(j)} \leq t} \left(\frac{n-j}{n-j+1}\right)^{1-\Delta_j} & \text{if } t < Y_{(n)} \\ 0 & \text{if } t \geq Y_{(n)} \end{cases} \qquad (1.15)$$

and obtain an estimator of the density function f as

$$f_n^*(t) = [nG_n^*(t)]^{-1} \sum_{i=1}^n k\left(\frac{t - Y_i}{h}\right)\delta_i.$$

The hazard rate can then be estimated as

$$h_n^*(t) = \frac{f_n^*(t)}{\widehat{S}_n(t)}.$$

We can replace the KM estimator $\widehat{S}_n(t)$ by any other reasonable survival function estimator in the denominator of the above estimator.

   The other popular failure rate estimators are the convolution-type estimators. The common form of a convolution estimator is

$$\hat{\lambda}_n(t) = \int \frac{1}{h} k\left(\frac{u - t}{h}\right) d\widehat{\Lambda}_n(u), t > 0 \qquad (1.16)$$

where $\widehat{\Lambda}_n$ is an estimator of the cumulative hazard function and k and h are a kernel function and bandwidth, respectively. One of the most common estimators used for Λ here is the Nelson–Aalen estimator $\widehat{\Lambda}$ defined in (1.9) above. One can replace $\widehat{\Lambda}$ by any suitable estimator of the cumulative hazard function Λ and obtain a smooth estimator of the failure rate. Due to the relative ease in obtaining its asymptotic properties and the exact mean squared error (see [25]), the convolution estimator has become the more popular between the two types of hazard rate estimators. A simple rearrangement shows that the above convolution estimator can be written as

$$\hat{\lambda}_n(t) = \sum_{i=1}^n \frac{\Delta_i}{(n-i+1)h} k\left(\frac{Y_{(i)} - t}{h}\right). \qquad (1.17)$$

For $0 < F(t) < 1$, both Tannner and Wong [32] and Yandell [35] derived the exact bias and the exact variance of $\hat{\lambda}_n(t)$ under slightly different conditions on the kernel function and the underlying distribution. For a large sample size n, it can be seen that

$$E[\hat{\lambda}_n(t)] - \lambda(t) = \frac{h^2}{2}\kappa_2\lambda''(t) + o(h^2)$$

and

$$\mathrm{Var}[\hat{\lambda}_n(t)] = \frac{\eta_2}{nh}\lambda(t)[1 - H(t)]^{-1} + o((nh)^{-1})$$

where κ_2 and η_2 are defined above and H is the cumulative distribution function of a Y. Thus, the mean squared error $E[\hat{\lambda}_n(t) - \lambda(t)]^2$ has the same type of leading terms and order terms as the mean squared error in density estimation for large sample sizes.

Singpurwalla and Wong [31] discuss the generalized jack-knife estimation of the hazard rate function with complete samples. The goal in their approach is to use kernels that can take negative values, and by an iterative estimation scheme, the mean squared error of the kernel density and hazard function estimator can be reduced. Toward the same goal, [20] discuss the use of kernels that can take negative values to reduce the bias of the above estimator when there is right censoring. In particular, by using kernels that satisfy

$$\int u^k k(u)\mathrm{d}u = 0, \quad k = 1, \ldots, m$$

for some $m > 2$, they discuss the bias reduction in the convolution-type estimator. However, the use of these kernels can lead to negative failure rate estimators. This is an undesirable feature. Marron and Wand [19] discuss the practical benefit (or lack thereof) of using a higher-order kernel. They point out that the gains from a higher-order kernel can be insignificant in most situations.

It is noteworthy that all smoothed estimators defined so far have a global smoothing parameter h that has to be chosen by some mechanism. We shall discuss various smoothing parameter selection procedures in the sequel. Tanner [32] considers a smooth kernel estimator of the failure rate using local smoothing in the following manner. Define R_r as the distance from the point t to the rth nearest failure time (i.e., to the rth nearest Y_i with the corresponding $\delta_i = 1$). Then, Tanner defines an estimator of the failure rate $\lambda(t)$ as

$$\tilde{\lambda}_n(t) = \frac{1}{2R_r}\sum_{i=1}^{n}\frac{\Delta_i}{n-i+1}r\left(\frac{Y_{(i)} - t}{2R_r}\right). \tag{1.18}$$

This estimator has a "smoothing parameter" R_r that is local to each t. Thus, the smoothness of the estimator adapts to the denseness/sparseness of the observed failure times near the point of estimation t. However, the choice of r decides how many adjacent lifetimes are used in the estimator. Hence, r here decides the smoothing parameter for a local smoothing window.

1.5.1 Local Linear Estimation of the Failure Rate

Local linear estimation of smooth functions has gained a substantial popularity in the recent years. In general, the local linear (or polynomial) estimators are based on local approximations to the unknown underlying function that is being estimated. For example, suppose a function g is being estimated at a point x. Then, the observations that are around x play a much more important role in estimating g at x than those away from x. Hence, we approximate g in a form

$$g(X) \approx g(x) + (X - x)g'(x)$$

and consider estimating the coefficients a and b locally in a linear form $a + b(X - x)$. This results in estimators of g (the estimator of a) and its derivative $g'(x)$ (the estimator of b). These ideas has been exploited in many smooth estimation problems including nonparametric regression, nonparametric quantile estimation, nonparametric density and failure rate estimation. In particular, in the case of failure rate estimation [10], propose a local linear estimator in the following fashion. Consider the optimization problem

$$\inf_{a_0, a_1} \int_{u \geq 0} \frac{1}{h} k\left(\frac{u - t}{h}\right) [\lambda(u) - (a_0 + a_1(u - t))]^2 du. \tag{1.19}$$

Then, the solution (a_0^*, a_1^*) to the above optimization problem approximates $(\lambda(t), \lambda'(t))$. Now, define

$$\lambda_n(t) = \sum_{i=1}^{n} \frac{\Delta_i}{(n - i + 1)} D(t - Y_{(i)})$$

where $D(t)$ is the Dirac delta function with the property that

$$\int g(u) D(u - x) du = g(x).$$

Then, it follows that $\int \lambda_n(t) dt = \hat{\Lambda}(t)$. Hence, it is reasonable to replace λ in (1.19) by λ_n to get

$$(\hat{a}_0, \hat{a}_1)' = \arg\inf_{a_0, a_1} \int_{u \geq 0} \frac{1}{h} k\left(\frac{u - t}{h}\right) [\lambda_n(u) - (a_0 + a_1(u - t))]^2 du \tag{1.20}$$

where $\hat{a}_1$ is the local linear estimator of $\lambda(t)$. In fact, a simple calculation shows that the estimators $\hat{a}_1$, $\hat{a}_2$ are the solutions to the linear equations

$$\sum_{i=1}^{n} \frac{1}{h} k\left(\frac{Y_i - t}{h}\right) [Y_{(i)} - t]^l \frac{\Delta_i}{(n - i + 1)} = \sum_{j=0}^{1} a_j \int_{u \geq 0} (u - t)^{j+l} \frac{1}{h} k\left(\frac{u - t}{h}\right) du$$

$$(1.21)$$

for $l = 0, 1$. If one examines the minimization of

$$\int_{u \geq 0} \frac{1}{h} k\left(\frac{u - t}{h}\right) [\lambda_n(u) - a_0]^2 du$$

with respect to a_0 for an interior point t in $[h, \tau)$, it is seen that

$$\hat{a}_0 = \sum_{i=1}^{n} \frac{\Delta_i}{h(n - i + 1)} k\left(\frac{Y_{(i)} - t}{h}\right),$$

which is the Tanner and Wong [33] estimator. Muller and Wang [20], 1994 have also studied this estimator. As mentioned before, the classical Tanner and Wong type estimators have boundary problems for estimating λ for $t \in [0, h)$. An advantage of using the local linear estimation is that the estimator $\hat{a}_0$ does not have boundary problems in the sense of a reduced order of the bias at the boundary. See [10] and [9] for a detailed discussion and some simulation examples, respectively.

1.5.2 Spline Estimation of the Failure Rate

A popular method in nonparametric estimation of smooth curves is the spline estimation. In a typical spline estimation, one tries to minimize a penalized distance or maximize a penalized likelihood where the penalty is imposed on the roughness of the curve being estimated. In estimating a failure rate, the penalized likelihood method can be described as follows.

For a right-censored sample with observations (Y_i, δ_i), $i = 1, \ldots, n$, the likelihood function is given by

$$L(\lambda) = \prod_{i=1}^{n} f(Y_i)^{\delta_i} S(Y_i)^{1-\delta_i}.$$

Then, using the relationship between λ and f, the log likelihood function of the censored data is given by

$$l(\lambda) = \sum_{i=1}^{n} \delta_i \ln(\lambda(Y_i)) - \int_0^{Y_i} \lambda(t) dt.$$

If no restrictions are imposed on the failure rate function λ, the above log likelihood is unbounded and hence the maximization of l with respect to λ does not yield a reasonable estimator of λ. Hence, a roughness penalty $J(\lambda)$ on the failure rate is incorporated into the above log likelihood resulting in a penalized log likelihood

$$\tilde{l}(\lambda) = \sum_{i=1}^{n} \delta_i \ln(\lambda(Y_i)) - \int_0^{Y_i} \lambda(t)\mathrm{d}t - \gamma J(\lambda).$$

Now, we can maximize $\tilde{l}(\lambda)$ within a class of functions(typically a Hilbert space) with respect to λ. Here, the parameter γ is a penalty factor that typically controls the penalization. A smaller value of γ produces a better fit (small bias) although a higher variability (rougher curve) of the estimator. A very common penalty is $J(\lambda) = \int [\lambda^{(2)}(t)]^2 \mathrm{d}t$ where $\lambda^{(2)}$ is the second derivative of the failure rate function λ. The role of γ above is very similar to the role of a smoothing parameter in kernel and local linear estimation. The minimization with a penalty of this type results in an estimator $\hat{\lambda}$ which is a cubic spline with knots at each $Y_{(i)}$. That is, the estimator $\hat{\lambda}$ is a piecewise cubic polynomial that has two continuous derivatives. If one uses the penalty $\int [\lambda'(t)]^2 \mathrm{d}t$ where λ' is the derivative of the failure rate function λ, the resulting estimator is a quadratic spline with knots at each $Y_{(i)}$ as shown by Anderson and Senthiselvan [2]. Although the solution of the estimator with the latter penalty is computationally simpler, a disadvantage of using this penalty is that the estimator can be negative between the observed failure times.

A detailed approach to estimate the hazard function via splines is given in Kooperberg, Stone and Young (1995). Suppose $\mathcal{G}$ be the p-dimensional linear space of functions on $[0, \infty)$ with a basis function $g_i(t), i = 1, \ldots, p$. Now, suppose we express the log-hazard rate $\log(\lambda(t)) = \sum_{i=1}^{p} \beta_i g_i(t)$. Then, we can write the log-likelihood function as

$$l(\beta_1, \ldots, \beta_p) = \sum_{i=1}^{n} \left[\delta_i \sum_{j=1}^{p} \beta_j g_j(Y_i) - \int_0^{Y_i} e^{\sum_{j=1}^{p} \beta_i g_i(t)} \mathrm{d}t \right]. \tag{1.22}$$

Letting

$$\psi(\beta_1, \ldots, \beta_p, Y_i, \delta_i) = \left[\delta_i \sum_{j=1}^{p} \beta_j g_j(Y_i) - \int_0^{Y_i} e^{\sum_{j=1}^{p} \beta_i g_i(t)} \mathrm{d}t \right],$$

the likelihood equations are

$$\frac{\partial l}{\partial \beta_j} = \sum_{i=1}^{n} \frac{\partial \psi(\beta_1, \ldots, \beta_p, Y_i, \delta_i)}{\partial \beta_j} = 0; \quad 1 \leq j \leq p \tag{1.23}$$

where

$$\frac{\partial \psi(\beta_1, \ldots, \beta_p, Y, \delta)}{\partial \beta_j} = \delta g_j(Y) - \int_0^Y g_j(t) e^{\sum_{j=1}^p \beta_i g_i(t)} dt.$$

Furthermore, the second-derivative matrix for the above log likelihood is given by

$$\frac{\partial^2 l}{\partial \beta_j \beta_k} = -\sum_{i=1}^n \int_0^{Y_i} g_j(t) g_k(t) e^{\sum_{j=1}^p \beta_i g_i(t)} dt; \quad 1 \le j, k \le p.$$

Noting that the log likelihood function is concave in the βs (Cox and Oakes, 1984), let the maximization of $l(\beta_1, \ldots, \beta_p)$ with respect to $\beta_i, i = 1, \ldots, p$ results in estimated coefficients $\widehat{\beta}_i, i = 1, \ldots, p$. This results in an estimator of the hazard rate

$$\hat{\lambda}(t) = e^{\sum_{j=1}^p \widehat{\beta}_j g_j(t)}.$$

The choice of the set $\mathcal{G}$ is typically the cubic splines where the optimal dimension p of the class $\mathcal{G}$ is obtained via a Bayesian information criterion. A very good discussion on kernel smoothing, local linear smoothing and spline smoothing can be found in Eubank [5].

1.6   Asymmetric Kernels

It is noteworthy that the lifetime data are all positive and that the underlying density function and the failure rate function are both supported on the positive half of the real line. This restriction inherits a spillover problem when attempting to estimate f and λ near the lower boundary (typically the origin). That is, when we use a symmetric kernel function k to estimate the density function, the estimated density $f(t)$ can take positive values for some negative values of t. The same issue prevails if one were to estimate the hazard rate using a symmetric kernel function. There are two approaches to avoid the spillover problem. The first is to use a boundary-adjusted kernel and the second is to use asymmetric kernels.

The boundary-adjusted kernels are varying kernels near the boundary of the observations. For example, if we are interested in estimating $\lambda(t)$ for $0 < t < h$ where h is the bandwidth, one can use a kernel of type k_t that satisfies $\int k_t(s) ds = 1$ and define an estimator

$$\hat{\lambda}(t) = \int \frac{1}{h(t)} k_t \left(\frac{t - x}{h(t)} \right) d\hat{\Lambda}(x)$$

where $h(t)$ is a bandwidth that depends on t. The choices of the local kernels k_t and local bandwidths $h(t)$ are discussed in Muller and Wang (1994).

The second approach to rectify the spillover issue in density estimation by using asymmetric kernels has been first attempted by Chen (1999) where a beta kernel was used to estimate densities with finite support. As a follow-up, [4] extended the density estimation to the positive real axis using a gamma kernel. A similar attempt was due to Scaillet [27] using an inverse Gaussian kernel. In all these attempts, there was a fundamental deficiency; the manner in which these authors have used the observations has lead to density estimators that did not integrate to 1 for a given bandwidth. Kulasekera and Padgett [14] proposed the use of inverse Gaussian kernels to estimate the density function with censored data where they remedied the issues associated with the estimators by previous authors. Kuruwita et al. [14] have unified the asymmetric kernel use to cover many candidate kernel functions such as gamma, inverse Gaussian, log normal, Weibull, etc. In their formulation, one writes the estimator of the density as

$$\hat{f}_n(t) = \int_0^\infty k(t, \psi(u, h), g(h)) d\hat{F}_n(u) \tag{1.24}$$

where the two functions ψ and g provide a reparametrization of the kernel density function to have the observations located at a point called the pivot; i.e., $\psi(u, h) = Y_{(i)}$ for each observation $Y_{(i)}$ and as $h \to 0$ the function $k(t, \psi(u, h), g(h))$ will become degenerated at each observation. Here, $\hat{F}_n$ is the estimated distribution function resulting from the KM estimator of the survival function as defined above. Since we choose k to be a proper probability density function, it then follows that

$$\int_0^\infty \hat{f}_n(t) dt = 1.$$

For example, suppose we use a log normal kernel function of type

$$k(x, \mu, h) = \frac{1}{\sqrt{2\pi}} \frac{1}{xh} e^{-\frac{1}{2} \frac{[\ln(x) - \ln(\mu)]^2}{h^2}}$$

where h is the bandwidth. If we position each observation Y_i at the median of the kernel function, we get $\psi(X, h) = \ln(X)$ and the estimator is of the form

$$\hat{f}_n(t) = \sum_{i=1}^n s_i \frac{1}{\sqrt{2\pi}} \frac{1}{th} e^{-\frac{1}{2} \frac{[\ln(t) - \ln(Y_{(i)})]^2}{h^2}}$$

where s_is are defined in (1.11). Now, notice that the integral of each summand over $[0, \infty)$ is 1 because each term is a log normal pdf with parameters $Y_{(i)}$ and h. Thus, the density estimator integrates to unity due to the property that $\sum s_i = 1$. This type of a representation is not unique, and for many asymmetric kernels, one can find many pivots that can give proper density estimators. However, for most pivots, the calculations are algebraically complex. We see in the sequel that the choice of the pivot becomes somewhat crucial in bandwidth selection.

For a detailed treatment of asymmetric kernels and the behavior of the resulting density estimators, we refer the reader to Kuruwita et al. [15]. Given the estimator of the density function, now one can develop a ratio-type estimator of the failure rate by constructing an estimator of the survival function based on $\hat{f}_n(t)$ as described in (1.4).

1.7 Bandwidth Selection

In all estimation problems discussed above, the nonparametric estimators involve user-defined bandwidths (smoothing parameters). The bandwidths that give asymptotic optimal rates for the mean squared error balance out the convergence rates of the bias and variance terms to zero as the sample size tends to infinity. For example, in the density estimation (and hence the failure rate estimation) with complete samples, it was seen that the bias of a regular second-order kernel estimator of the density is proportional to h^2 and the variance is proportional to $1/nh$ for a bandwidth h. Since for the kernel estimator to be mean square consistent, it will be required that the bandwidth parameter to converge to 0 while $nh \rightarrow \infty$ as the sample size increases to ∞. Thus, we see that a rate $n^{-1/5}$ for h balances the convergence rate of the squared bias and the variance of the nonparametric density estimator. It has been shown that the same asymptotic rates remain valid for mean squared error optimal hazard rate estimation. In both these estimation problems (whether the sample is complete or censored) when the bandwidth is selected to be of order $n^{-1/5}$, the mean squared error converges at a rate of $n^{-4/5}$. This rate is referred to as the optimal second-order rate in kernel estimation, and it is slower than the classical n^{-1} rate for the mean square rate in regular parametric estimation problems (Lehmann and Casella, 1998).

Although bandwidth parameters of type $Cn^{-1/5}$ where C is a constant give the optimal rate of convergence for the pointwise and integrated mean squared error for nonparametric density and hazard rate estimation, the behavior of kernel estimators can drastically change when one changes the constant C in finite samples. As seen above, in kernel density and hazard rate estimation, the constant C depends on the underlying density function and the hazard rate function. For example, in density estimation with complete samples, the leading terms in the pointwise mean squared error $E[f_n(t) - f(t)]^2$ are

$$M(h, t) = \frac{1}{4}h^4 \kappa_2^2 (f''(t))^2 + \frac{1}{(nh)}\eta_2$$

where κ_2 and η are determined by the kernel function. Now, the optimal bandwidth h_0 that minimizes $M(h)$ with respect to h is given by

$$h_0 = \left[\frac{\eta_2}{n(\kappa_2 f''(t))^2} \right]^{-\frac{1}{5}}. \tag{1.25}$$

Hence, the determination of the exact optimal bandwidth requires the knowledge of f'' at the point of estimation. If we are using the optimal integrated mean squared error defined as

$$\text{MISE}(f) = \int E[f_n(t) - f(t)]^2 w(t) dt$$

for a suitable weight function w as a precision guideline, then we can minimize $\int M(h, t) dt$ with respect to h and obtain an optimal bandwidth h_0^*, which is a global bandwidth for estimating the density function at any t. The form of this optimal bandwidth is easily obtained by a similar argument as above, and we see that h_0^* depends on $\int [f''(t)]^2 dt$, again an unknown quantity. Thus, in both cases above, to obtain a working bandwidth, we have to plug in estimators of either f'' or an integrated version of it into the bandwidth formula. Such bandwidths are thus called "plug-in" bandwidths.

In density and hazard rate estimation with censored observations, the pointwise mean squared error and the integrated mean squared error of non parametric estimators have very similar expressions as those in complete sample case. As seen in Sect. 1.5, the mean squared error of estimating the hazard rate using a ratio-type kernel estimator is given by

$$\frac{h^4}{4} \kappa_2^2 [\lambda''(t)]^2 + \frac{\eta_2}{nh} \lambda(t)[1 - H(t)]^{-1} + o(h^2) + o((nh)^{-1})$$

where H is the cumulative distribution of the observation $Y = \min(T, L)$. Hence, the leading terms of the pointwise mean squared error in this case are given by

$$\tilde{M}(t) = \frac{h^4}{4} \kappa_2^2 [\lambda''(t)]^2 + \frac{\eta_2}{nh} \lambda(t)[1 - H(t)]^{-1},$$

and the bandwidth that optimizes $\tilde{M}(t)$ is of the same order as in the case of density estimation with complete samples. Then, the bandwidth that yields the optimal pointwise MSE (in its leading terms) is

$$h^* = \left[\frac{\lambda(t)\eta_2}{n(1 - H(t))[\kappa_2 \lambda''(t)]^2} \right]^{\frac{1}{5}}.$$

In this case, for obtaining a plug-in bandwidth, one needs estimators of the underlying density function (or the distribution function) and the distribution $(1 - G)$ of the censoring variable L.

An alternative and popular method is the least square cross-validation (CV). The basic idea in CV is to write a near-unbiased estimator of the mean squared error or the integrated mean squared error using a leave-one-out estimator of the underlying density or the hazard rate function using a given bandwidth h. This MSE of the estimator is then minimized with respect to h to obtain an adaptive bandwidth that has been shown to work well in general. We shall describe the CV [23] method in some detail for estimating either the density or the hazard rate function in the following. We start by generalizing the smooth estimation problem under censoring. To accomplish this, we start by defining a function ξ

$$\xi(t) = \frac{(1 - G(t))f(t)}{Q(t)}$$

and consider estimating this function with an adaptive bandwidth. Here, $Q(t)$ is a nondecreasing function such that $0 \le Q(t) \le 1$ for all real numbers t. When $Q(t) = 1 - G(t)$, the ξ function becomes the density function, and when $Q(t) = 1 - H(t)$, the ξ is the hazard rate function. Now, we define the empirical distribution function of the Y_is as

$$H_n(t) = \frac{1}{n} \sum_{i=1}^{n} I[Y_i \le t].$$

Now, we define the estimator G_n via the equality $1 - H_n = S_n(t)[1 - G_n(t)]$ where S_n is the Kaplan–Meier estimator of S. Then, define the estimator $\xi_n(t)$ of $\xi(t)$ by

$$\xi_n(t) = \frac{1}{n} \sum_{i=1}^{n} \frac{k\left(\frac{Y_i - t}{h}\right) \Delta_i}{Q_n(Y_i)} \tag{1.26}$$

where Q_n is an estimator of the function Q. The estimator of Q can be $1 - H_n$ if we are estimating the hazard rate. If we are estimating the density f, then we can get an estimator of Q $(1 - G)$ by reversing the definition of the censoring for T and L and using the KM estimator. Now, further define

$$\bar{\xi}(t) = \frac{1}{n} \sum_{i=1}^{n} \frac{k\left(\frac{Y_i - t}{h}\right) \Delta_i}{Q(Y_i)}.$$

It is easy to see that

$$|\xi_n(t) - \bar{\xi}(t)| = O_p(n^\alpha)$$

if $|Q_n(t) - Q(t)| = O_p(n^\alpha)$. The optimal bandwidth is selected to minimize a sample version of the integrated squared error (ISE) defined as

$$\Gamma(h, \xi_n) = \int [\xi_n(t) - \xi(t)]^2 w(t) dt$$

where w is a suitable weight function. Since, under some regularity conditions on interchanging the expectation and integration, $E[\Gamma(h, \xi_n)] = \text{MISE}(\xi, h)$, we expect the minimizer of $\text{MISE}(\xi)$ to provide a reasonable bandwidth for estimating ξ. If we knew the function ξ, then the bandwidth that minimizes $\text{MISE}(\xi, h)$ gives an optimal bandwidth in an average sense over all possible data sets. However, minimizing $\text{MISE}(\xi, h)$ is not practical because it involves the unknown ξ in it. Now, considering the sample at hand, we attempt to circumvent this issue by proposing to minimize a suitable sample version of $\Gamma(h, \xi_n)$ with respect to h as an approximation to the minimizer of $\text{MISE}(\xi, h)$ with respect to h. Such a sample version of $\Gamma(h, \xi_n)$ is generally referred to as a cross-validator ($CV(h)$). In this case, we can define $CV(h)$ in the following manner. Note that

$$\Gamma(h, \xi_n) = \int \left[\xi_n^2(t)w(t) - 2w(t)\xi(t)\xi_n(t) + \xi^2(t)w(t) \right] dt.$$

The last term on right side does not involve h, and therefore, we can minimize the sum of the first two terms to obtain an optimal h. The integral of the second term can be nearly unbiasedly estimated by

$$\frac{1}{n} \sum_{i=1}^{n} \frac{\xi_{i,n}(Y_i)}{Q_n(Y_i)} w(Y_i)\Delta_i$$

where $\xi_{i,n}(t)$ is the leave-one-out estimator of ξ defined as

$$\xi_{i,n}(t) = \frac{1}{n} \sum_{j \neq i}^{n} \frac{k\left(\frac{Y_j - t}{h}\right)\Delta_j}{Q_n(Y_j)}.$$

Hence, the bandwidth $\hat{h}$ that minimizes

$$CV(h) = \int \left[\xi_n^2(t)w(t)dt \right] - \frac{1}{n} \sum_{i=1}^{n} \frac{\xi_{i,n}(Y_i)}{Q_n(Y_i)} w(Y_i)\Delta_i$$

is considered an optimal adaptive bandwidth for estimating ξ. For a weight function $w(t) = 1$ for all $t \in [0, T]$ where $T = \sup\{x | Q(x) > \epsilon\}, \epsilon > 0$, it has been shown [18, 23] that the above bandwidth is asymptotically equivalent to the minimizer h_0 of $\text{MISE}(\xi, h)$. In particular, it has been proven that

$$\frac{\hat{h} - h_0}{\hat{h}} = O_p(n^{-1/10})$$

as $n \to \infty$ for a second-order kernel.

To illustrate the smooth estimation of the failure rate, we generated 200 data points from a distribution with a linear hazard rate $\lambda(x) = 2t$, which are censored by exponential censoring variables with mean 1. This resulted in 95 censored values. we then applied the convolution method and the cubic spline method to estimate the hazard rate where the bandwidth is chosen by the cross-validation

Fig. 1.3 Plots of failure rate estimators for Insulation Fluid Data. *Hard Line* Kernel estimator; *Broken Line* linear spline estimator; *Dotted Line* cubic spline estimator

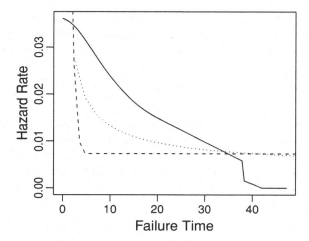

discussed above. The plots in Fig. 1.1 gives the convolution-type kernel estimator (hard line) and the cubic spline–smoothed hazard rate (broken line). We note that the two estimators for this sample produce almost identical failure rate estimators, which seem to be reasonably accurate (the true rate being $2t$).

The results of application of the kernel estimator, the linear spline estimator and the cubic spline estimator to estimate the failure rate function for the fluid insulation data are given in Fig. 1.3. The data reported in Lawless [16] have 26 observations with 11 censored values. Although there are various temperature values that were measured along with the breakdown data, we ignore the covariate effect here and model the data as a single sample. All estimators indicate a decreasing failure rate function, and the cubic spline method produces the smoothest estimator where all the bandwidths have been selected by a cross-validation.

1.7.1 Bayesian Bandwidths

One main issue in density and hazard estimation with lifetime data is the spillover problem mentioned above. As discussed, one remedy is to use an asymmetric kernel function. Along with asymmetric kernels, the pointwise bandwidth can be chosen by a Bayesian-type argument [14, 15]. The basic approach is to consider the bandwidth h as a parameter that needs to be estimated using a Bayes argument. Here, we briefly discuss the density estimation with censored data. Consider the function $f_h(t)$ defined as

$$f_h(t) = \int k_h(x - t) \mathrm{d}F(x)$$

where $k_h(u) = \frac{1}{h} k(u/h)$.

Then, $\hat{f}_h$ given in (1.10) is a natural estimator of f_h where we use $1 - \hat{S}_n$ as an estimator of F. Now, suppose we use an inverse Gaussian (IG) kernel k so that the 'spillover' problem at the origin is prevented (see [4, 27, 30]); i.e., we use

$$k(x; \mu, h) = \frac{1}{\sqrt{2\pi h x^3}} e^{-\frac{1}{2h}\frac{(x-\mu)^2}{x\mu^2}}, \quad x > 0,$$

where $\mu, h > 0$ are the mean and the scale parameter, respectively. Then, $\hat{f}_h$ is a proper density for each n, and the smoothing parameter h is the scale parameter for the IG density. We explore the ideas from Bayes estimation to find a suitable value for h given the data, the observations and censoring indicators. Hence, for a prior density $\xi(h)$, the posterior density of h at the point of estimation x is

$$\xi(h|x) = \frac{f_h(x)\xi(h)}{\int f_h(x)\xi(h)\mathrm{d}h}.$$

However, since we do not know f_h, given $\mathcal{X} = \{(Z_i, \Lambda_i), i = 1, \ldots, n\}$ (the data), reminiscent of an empirical Bayes approach with one data observation step, we can estimate the posterior by

$$\hat{\xi}(h|x, \mathcal{X}) = \frac{\hat{f}_h(x)\xi(h)}{\int \hat{f}_h(x)\xi(h)\mathrm{d}h}.$$

Then, for the squared error loss, an estimator of h is given by the estimated posterior mean

$$\tilde{h}(x) = \int h\hat{\xi}(h|x, \mathcal{X})\mathrm{d}h. \tag{1.27}$$

Note that with this approach, the posterior is a function only of h and, with a simple prior structure, $\hat{\xi}$ and $\tilde{h}$ can be explicitly obtained.

Suppose we use an inverted gamma prior with parameters α and β and density

$$\xi(h) = \frac{1}{\beta^\alpha \Gamma(\alpha) h^{\alpha+1}} e^{-\frac{1}{\beta h}}.$$

Then straightforward, but tedious, calculations give the (estimated) posterior density

$$\hat{\xi}(h|x, (Z_i, \Lambda_i), i = 1, \ldots, n) = \frac{\sum_{j=1}^n \frac{s_j}{Z_j^{3/2} h^{\alpha^*+1}} e^{-\frac{1}{\beta_j^* h}}}{\Gamma(\alpha^*) \sum_{j=1}^n \frac{s_j (\beta_j^*)^{\alpha^*}}{Z_j^{3/2}}}$$

where $\alpha^* = \alpha + \frac{1}{2}$ for $\alpha > 1/2$ and

$$\beta_i^* = \left[\frac{1}{\beta} + \frac{(x - Z_i)^2}{2xZ_i^2}\right]^{-1}.$$

Then, the Bayes estimator of h at x is found as the mean of this posterior,

$$\tilde{h}(x) = \frac{\sum_{i=1}^n s_i \beta_i^{*(\alpha^*-1)}}{(\alpha^* - 1) \sum_{i=1}^n s_i \beta_i^{*\alpha^*}}. \tag{1.28}$$

If we were to use an inverted gamma density with a finite mean, then we may choose $\alpha > 1$ in the previous calculation. If we were to use the improper prior

$$\xi(h) \propto \frac{1}{h^2},$$

then the proper posterior density can be found similarly as

$$\hat{\xi}(h|x, (Z_i, \Lambda_i), i = 1, \ldots, n) = \frac{2}{\sqrt{\pi}} \frac{\sum_{j=1}^n s_j h^{-2/5} \exp[-1/(\beta_j^{**}h)]}{\sum_{j=1}^n s_j (\beta_j^{**})^{3/2}}$$

and the resulting estimator of h is

$$\tilde{h}(x) = 2 \frac{\sum_{i=1}^n s_i \beta_i^{**1/2}}{\sum_{i=1}^n s_i \beta_i^{**3/2}},$$

where β_i^{**} is given by

$$\beta_i^{**} = \left[\frac{(x - Z_i)^2}{2xZ_i^2}\right]^{-1}.$$

One can then use these Bayesian local bandwidths in estimating the density using (1.10) at a given x for any sample size n. Note that $\tilde{h}(x)$ can be calculated using only the observed data, and prior parameters in the proper prior case. The classical asymptotically optimal bandwidth involves the unknown density itself, which must be estimated for bandwidth calculations. In the approach given here, the Bayesian bandwidth is exact for all sample sizes, and there is nothing to estimate for calculating the bandwidth.

It is clear that the behavior of $\tilde{h}$ then primarily depends on the behavior of the prior density near the origin. By judicious choices of prior parameter values of the prior distribution, [14] show that this bandwidth h approaches zero as $n \to \infty$ as required for the consistency of estimation of the failure rate. A generalization to other kernels and priors is discussed with several examples in Kuruwita et al. [15].

References

1. Aalen O (1978) Non-parametric estimation of partial transition probabilities in multiple decrement models. Ann Stat 6:534–545
2. Anderson JA, Senthiselvan A (1980) Smooth estimates for the Hazard function. J R Stat Soc (B) 3:322–327
3. Breslow N (1972) Discussion on professor Cox's paper, introduction to stochastic processes in biostatistics. Wiley, New York
4. Chen SX (2000) Probability density function estimation using gamma kernels. Ann Inst Stat Math 52:471–480
5. Eubank RL (1999) Nonparametric regression and spline smoothing. Marcel Dekker, New York
6. Fleming TR, Harrington DP (1991) Counting processes and survival analysis. Wiley, NY
7. Gasser T, Muller HG (1979) Kernel estimation of regression functions. It smoothing techniques for curve estimation 23–68
8. Ghorai JK, Susarla V (1990) Kernel estimation of a smooth distribution function based on censored data. Metrika 37:71–86
9. Hess KR, Serachitopol BM, Brown DW (1999) Hazard functions estimators: a simulation study. Stat Med 18:3075–3088
10. Jiang J, Doksum K (2003) On local polynomial estimation of hazard rates and their derivatives under random censoring, IMS Lecture Notes-Monograph Series, vol 42, Mathematical Statistics and Applications. pp 463–481
11. Kaplan EL, Meier P (1958) Nonparametric estimation from incomplete observations. J Am Stat Assoc 53:457–481
12. Karunamuni R, Yang S (1991) Weak and strong uniform consistency rates of kernel density estimates from randomly censored data. Can J Stat 19:349–359
13. Kulasekera KB, Williams CL, Manatunga A (2001) Smooth estimation of the reliability function. Lifetime Data Anal 415–433
14. Kulasekera KB, Padgett WJ (2006) Bayes bandwidth selection in kernel density estimation from censored data. J Nonparametr Stat 18:129–143
15. Kuruwita CN, Kulasekera KB, Padgett WJ (2010) On boundary bias and bandwidth selection in density estimation with censored data, to appear in the J Stat Plan Inference
16. Lawless JF (1998) Statistical models and methods for lifetime data. Wiley, NY
17. Lehman LE, Cassella G (1998) Theory of point estimation. Springer, Berlin
18. Marron JS, Padgett WJ (1987) Asymptotically optimal bandwidth selection for kernel density estimators from randomly right-censores samples. Ann Math Stat 15:1520–1535
19. Marron JS, Wand MP (1992) Exact mean integrated squared error. Ann Stat 20:712–736
20. Muller HG, Wang JL (1990) Locally adaptive hazard smoothing. Prob Theory Related Fields 85:523–38
21. Nelson W (1984) Applied life data analysis. Wiley, NY
22. Parzen E (1962) On estimation of a probability density and mode. Ann Math Stat 35:1065–1076
23. Patil PN (1993) On the least squares cross-validation bandwidth in hazard rate estimation. Ann Stat 21:1792–1810
24. Prakasa Rao BLS (1983) Nonparametric functional estimation. Academic Press, Orlando
25. Rice J, Rosenblatt M (1976) Estimation of the log survivor function and hazard function. Sankhya 38:60–78
26. Reiss RD (1981) Nonparametric estimation of smooth distribution functions. Scand J Stat 8:116–119
27. Scaillet O (2004) Density estimation using inverse and reciprocal inverse Gaussian kernels. J Nonparametr Stat 16:217–266
28. Singh RS (1977) Improvement on some known nonparametric uniformly consistent estimators of derivatives of a density. Ann Stat 5:394–399

29. Schucany WR, Sommers JP (1977) Improvement of kernel type density estimation. J Am Stat Assoc 72:420–423
30. Silverman B (1986) Density estimation for statistics and data analysis. Chapman and Hall, London
31. Singpurwalla N, Wong WH (1983) Kernel estimators of the failure rate function and density estimation: an analogy. J Am Stat Assoc 78:478–481
32. Tanner M (1983) A note on the variable kernel estimator of the hazard function from randomly censored data. Ann Stat 11:994–998
33. Tanner M, Wong WH (1983) The estimation of the hazard function from randomly censored data by the kernel method. Ann Stat 11:989–993
34. Uzunogullari U, Wang J-L (1992) A comparison of hazard rate estimators for left truncated and right censored data. Biometrika 79:297–310
35. Yandell B (1983) Nonparametric inference for rates with censored survival data. Ann Stat 11:1119–1135

Part II
Grey-Box Approach: Counting Processes

Most reliability systems are assumed to evolve (deteriorate) somehow over time. Since maintenance decisions depend greatly on knowledge of the system conditions at any time, a model of the temporal uncertainty associated with the evolution of the performance properties of the system is needed. This suggests that some of the physical components in the modeling scheme should be considered, specifically time-dependence. Therefore, the choice of a black-box model such as those considered up till now may prove insufficient. In this part of the book we will consider various types of grey-box model. As was established by Nicolai (2008, Nicolai, R.P., Maintenance models for systems subject to measurable deterioration, Ph.D. thesis, Erasmus University, Rotterdam), we consider that a grey-box deterioration model is based on a measurable quantity indicating time-dependent deterioration and failure, i.e. a stochastic process. Unlike black-box models, the grey-box model provides a mathematical representation, but some of the physics is approximated. For our purposes, the model is reduced in order to indicate the time-dependent deterioration and failure. Because time dependence is introduced, a stochastic process is required to model the deterioration of a repairable system. We focus on this problem, paying some attention to the underlying maintenance policies in each case.

- Perfect repair. Once the system has failed and a repair has been completed, its behavior is exactly the same as if it were new. To model this situation, renewal processes are considered. Traditionally, within the scope of a non-parametric approach, empirical estimators have been constructed to estimate the main performance measures of such a system, i.e. renewal function, point availability and ROCOF. In addition, some recent works have tried to improve the accuracy of empirical estimators by constructing kernel-smooth versions.

- Minimal repair. In this situation, once the system has been repaired, its state is identical to that just before the failure. The typical modeling in this case is a non-homogenous Poisson process. In the past, the majority of estimation procedures performed for such a case were based on the assumption that the shape of the ROCOF fitted one of two possibilities: log-linear or power law, so there is an overwhelming list of publications with a parametric scope. However, we do not consider any functional form for the ROCOF but look at the results obtained by authors who use bootstrap techniques to construct (among other things) confidence bands for the ROCOF.

- Imperfect repair. In practice, a more general situation than those described in the two previous paragraphs may arise. After the failure has occurred and the repair action has been completed, it would be expected that the system be restored to a level state that is an improvement on the one presented by the system just before the failure. The state of the system is better than the one before the failure but not necessarily as good as new. Models to accommodate such features are in the literature usually called imperfect repair models. The classical such model is the so-called Brown–Proschan model, which has been generalized in several directions using the notion of effective age. Examples of such approaches, involving nonparametric estimation, will be considered. A different approach has recently been considered in the literature: the trend-renewal process, which contains both a renewal process and a non-homogeneous Poisson process as special cases. Very few research has been done until now with respect to the nonparametric estimation problem of the characteristics of such processes (e.g. the conditional intensity function). However, in some recent publications, interesting approaches have been developed and our study is based on these.

Chapter 2
Models for Perfect Repair

2.1 Introduction

According to the definition given by Ascher and Feingold [2], a repairable system is understood to be a system which, after failure, can be restored to a functioning condition by some maintenance action other than replacement of the entire system. Replacing the entire system may be an option, but it is not the only one. In this part of the book, we will assume that the description of the system state at any time is reduced to two levels: operative and failed. More detailed specifications of the state space are considered in Part III of the book.

Model deterioration (performance) of a repairable system can be tackled in several different ways. On the one hand, the interest may lie mainly in modeling the number of failures suffered by the system up to time t. If $N(t)$ is the number of failures of a repairable system occurring in the interval $(0, t]$, the most appropriate approach is to consider the counting process given by $\{N(t), t \geq 0\}$ as model deterioration. Attention is usually focused on the expected value, variance, and probability distribution of $N(t)$. The homogeneous Poisson process (*HPP*) is the counting process most frequently used throughout the extensive literature on the subject. *HPP* may also be characterized in terms of the random length of the times between two consecutive failures, exponentially distributed with the same parameter.

Data from repairable systems are usually given as times between failures $T_1, T_2, \ldots$. A common assumption made is that these failure times are independent and identically distributed, with distribution F. This assertion implies that after a failure, the system behavior is exactly the same as if it were new; thus, a perfect repair maintenance action is being carried out in the system environment. As explained in Kijima [26], in practice, the perfect repair assumption may be reasonable for systems with one structurally simple unit.

When F denotes a general family of distributions, the sequence $\{T_1, T_2, \ldots\}$ is referred to as a renewal process (*RP*). Therefore, the *HPP* may be seen as a

M. L. Gámiz et al., *Applied Nonparametric Statistics in Reliability*,
Springer Series in Reliability Engineering, DOI: 10.1007/978-0-85729-118-9_2,
© Springer-Verlag London Limited 2011

particular case of *RP*. Of course, there is an evident duality between "time domain" and "counts domain", i.e., $\{N(t) \geq k\}$ if and only if $\{T_1 + T_2 + \cdots + T_k \leq t\}$. Putting this into words, there have been at least k renewals until time t if and only if the kth renewal has occurred before t.

Another general assumption made when using a counting process to model the time-evolution of a repairable system is that the time-to-repair is negligibly small compared to its time-to-failure. In many practical applications, where it is reasonable to expect that the system is not under repair for long in relation to its operating time, this assumption is fairly realistic. Otherwise, the system is not feasible.[1] So, in a case where it is assumed that the system is repaired and put into new operation immediately after the failure, the deterioration model will be given by an Ordinary Renewal Process (*ORP*).

On the other hand, there are some situations where one is interested in estimating other important measures such as availability (unavailability) of the system, which is the probability that the system is functioning at a given time. In this case, the modeling tool indicated is the Alternating Renewal Processes (*ARP*) where operative periods alternate with repair periods. Within the scope of an *ARP*, data collected consist of a sequence of alternating lifetimes and repair times, i.e. (T_1, R_1), (T_2, R_2), ..., where T_1, T_2, ... are the successive lifetimes of the system and these are independent and identically distributed (*i.i.d.*) with *CDF F*; and R_1, R_2, ..., the corresponding repair times, which are i.i.d. with *CDF G*. It is also assumed that T_i and R_i are independent, for any $i = 1, 2, \ldots$. Every random length obtained as a lifetime plus a repair time is called a renewal cycle. Repairable system data are collected to estimate among other measures, quantities such as:

- The distribution of lifetimes (respectively, repair times);
- The expected number of renewals in an interval $(0, t]$, which is the renewal function;
- The probability that the system is operative at a given time t, which is the instantaneous availability;
- The proportion of time the system is in a functioning condition, which is the steady state or limiting availability.

Inference studies are carried out without assuming any particular functional form for distribution functions F and G. We therefore use a nonparametric approach. From a given data set, empirical estimators are constructed for the performance measures of a repairable system whose time evolution is modeled by an *ORP* or an *ARP* and we also obtain smooth estimators based on kernel functions. The implicit bandwidth parameter is derived by means of data-driven procedures, specifically bootstrap techniques, which prove very easy to implement and give very good results, as pointed out in the simulation examples included.

---

[1] Sometimes one may let "operating time" be the time parameter; or possibly "number of cycles" or "number of kilometers" (for cars). Then, repair times are 0 in these time axes.

2.2 Ordinary Renewal Process

In this section, we study probabilistic models for systems where after the occurrence of a random event (failure), at a random time, everything in the system starts over at the beginning. So, we consider systems under perfect repair maintenance policies, which means that the system operating state is restored to "as good as new" conditions after failure. This approach is appropriate for systems such as light bulbs or thermometers, where the occurrence of failures implies the substitution of the entire system, not being available partial repairs to recover the system function. Furthermore, if a reliability system is understood as a set of components or elementary parts, renewal processes are plausible models for the time behavior of the parts better than for the whole system, since after a failure occurs in a component, the replacement of the component is usually carried out, instead of repair.

Under perfect repair models, at time $t = 0$ a repairable system is put into operation and is functioning. At each failure time, the system is replaced by a new one of the same type. This process is repeated along time, and the replacement time is considered negligible. As a result, a sequence of lifetimes or random variables which are independent and identically distributed is obtained. Renewal processes have been extensively used by many researches interested in reliability (Barlow and Proschan [3] or Rausand and Hoyland [36] are classical references), the most simple case being the Homogenous Poisson Process (*HPP*), where the random time between successive renewals has an exponential distribution.

2.2.1 The Renewal Function

Renewal theory arises from the study of stochastic systems whose time evolution appear as successive life cycles. A life cycle is a time interval during which the system is functioning. At the start of every interval, the system is stochastically reinitiated. In this section, we introduce the main features that characterize an ordinary renewal process, paying special attention to the renewal function.

As stated above, an *ORP* may be represented by means of a sequence of random independent and identically distributed variables $\{T_k; k = 1, 2, \ldots\}$ (we will consider only non-negative variables) or equivalently by means of the counting process $\{N(t); t \geq 0\}$, where $N(t) = max\{k: T_1 + T_2 + \cdots + T_k \leq t\}$, that is, the number of renewals occurring in the interval $(0, t]$. Let F denote the *CDF* common to all T_k, and let us define $S_0 = 0$ and $S_k = T_1 + T_2 + \cdots + T_k$, for $k = 1, 2, \ldots$, as the so-called *waiting times*, making it obvious that $P\{N(t) \geq k\} = P\{S_k \leq t\}$. This random quantity, S_k is obtained as the sum of k independent random variables, so that its *CDF*, which we call F_k, is given by the k-fold *Stieltjes-convolution* of F for $k \geq 1$, that is,

$$F^{(k)} = P\{S_k \leq t\} = F^{(k-1)} * F(t) = \int_0^t F^{(k-1)}(t-u)dF(u).$$

The main objective in renewal theory is to derive the properties of $N(t)$, in particular its expected value, which is called the *renewal function*.

Definition 2.1 *(Renewal Function)* Let $F(t) = P\{T \leq t\}$ be the *CDF* of the lifetime of a repairable system with perfect and instantaneous repair. Let $\{N(t), t \geq 0\}$ be the corresponding renewal counting process. The renewal function is defined as

$$M(t) = E[N(t)], \quad \text{for } t \geq 0.$$

It is easy to check the following equalities

$$M(t) = E[N(t)] = \sum_{k=1}^\infty P\{N(t) \geq k\} = \sum_{k=1}^\infty P\{S_k \leq t\} = \sum_{k=1}^\infty F^{(k)}(t).$$

It can also be stated that $M(t) < \infty$ for all $0 \leq t < \infty$. Furthermore, the expression above may be given via the following integral representation

$$M(t) = F(t) + \int_0^t M(t-u)dF(u) = F(t) + M(t) * F(t), \quad \text{for } t \geq 0, \qquad (2.1)$$

which is a particular case of a wider class of equations called *renewal equations*,

$$W(t) = v(t) + \int_0^t W(t-u)dF(u) = v(t) + W(t) * F(t), \quad \text{for } t \geq 0,$$

where $v(t)$ and $F(t)$ are known, whereas $W(t)$ is an unknown function. In other words, $M(t)$ satisfies the renewal equation given by (2.1), and moreover, it is the unique solution that is bounded on finite intervals.

Closed form analytic expressions for $F^{(k)}$ are not generally available, special cases are Erlang and Normal distributions. Based on a central result of renewal theory, the **key renewal theorem** (there exists an extensive literature over the subject, see for instance [37]), simple asymptotic approximations can be obtained in the case where $E[T] = \mu < \infty$ and $Var[T] = \sigma^2 < \infty$,

$$\lim_{t \to \infty} \left[M(t) - \frac{t}{\mu} \right] = \frac{\sigma^2}{2\mu^2} - \frac{1}{2}.$$

This expression suggests the following asymptotic expression for the renewal function

$$M_\infty(t) = \frac{t}{\mu} + \frac{\sigma^2}{2\mu^2} - \frac{1}{2}.$$

With these considerations, the following estimator for the renewal function, valid for large values of t, may be defined,

$$\widehat{M}_\infty(t) = \frac{t}{\widehat{\mu}} + \frac{\widehat{\sigma}^2}{2\widehat{\mu}^2} - \frac{1}{2},$$

where $\widehat{\mu}$ and $\widehat{\sigma}$ are estimators of μ and σ respectively, based on data recorded up to time t. Nevertheless, in certain application areas such as reliability engineering, the interest is rather in the initial part of the life of a device, that is, for $t \in [0, 3\mu]$, see Frees [13] and Gertsbakh and Shpungin [17], where the estimation problem becomes more difficult. It seems natural to estimate the renewal function based on a sum of estimators of the convolutions of F, that is, we define

$$\widehat{M}(t) = \sum_{k=1}^{\kappa} \widehat{F}^{(k)}(t) \tag{2.2}$$

where the number of terms in the summation, the parameter κ, has to be determined. Various ways have been proposed by different authors in the literature on the subject.

2.2.2 Nonparametric Estimation of the k-Fold Convolution of Distribution Functions

The problem of dealing with the function M involves estimating k-fold convolution functions, which is not an easy task. Recently, a number of authors have tackled the problem, revealing the inherent difficulty in most cases. Below, we present some of these. Let $T_1, T_2, \ldots, T_n$ be non-negative independent random variables, with cumulative distribution function F. Let $F^{(k)}$ be the k-fold convolution function of F.

2.2.2.1 The Empirical Convolution Function

Frees [14] defines two alternative estimators of the renewal function. The first is based on the sum of the convolutions without replacing the empirical distribution function, and the second, called the *empirical renewal function*, is obtained as the renewal function of the empirical distribution corresponding to F. Let us introduce the estimators of the convolutions which Frees defines for constructing estimators of type $\widehat{M}(t)$ as given above.

If $\{i_1, i_2, ..., i_k\}$ is a subset of size k of $\{1, 2, ..., n\}$, then an estimator of $F^{(k)}(t)$ is

$$\widehat{F}_{C1}^{(k)}(t) = \frac{1}{\binom{n}{k}} \sum_{(n, k)} I(T_{i_1} + T_{i_2} + \cdots + T_{i_k} \leq t), \qquad (2.3)$$

where $\sum_{(n, k)}$ denotes the sum over all $\binom{n}{k}$ different index combinations $\{i_1, i_2, ..., i_k\}$ of length k. The estimator (2.3) is a U-statistic and therefore it can be established that, for each $k \geq 1$, and for each $t \geq 0$, that

$$\widehat{F}_{C1}^{(k)}(t) \longrightarrow F^{(k)}(t), \text{almost surely}(a.s.).$$

Moreover, it is an unbiased minimum-variance estimator of $F^{(k)}(t)$. Based on the estimator in (2.3), Frees obtained the uniform consistency of $\widehat{M}_{C1}(t)$ $a.s.$ in compact intervals $[0, t]$, on the assumption that the number of terms in (2.2), $\kappa = n$ and that T has a positive mean and finite variance. The asymptotic normality of $\widehat{M}_{C1}(t)$ is also proven under some moment conditions.

The drawback of this estimator is the considerable number of computations needed to evaluate it, even though Frees introduced the design parameter $\kappa \leq n$. Schneider et al. [39] propose a new algorithm to compute the Frees estimator in order to reduce the computation time. They define a family of characteristic functions based on the sample that can be determined recursively, and then use Fourier transforms to recover the distributions $\widehat{F}_{C1}^{(k)}(t)$.

An alternative estimator of $M(t)$ is defined in the Concluding Remarks section in Frees [14]. In this case, $F^{(k)}(t)$ is estimated by means of the k-fold convolution of the empirical distribution function obtained from $T_1, T_2, ..., T_n$,

$$\widehat{F}_{C2}^{(1)}(t) = \frac{1}{n} \sum_{i=1}^{n} I(T_i \leq t) \qquad (2.4)$$

for $k = 1$, which is the empirical distribution function, and

$$\widehat{F}_{C2}^{(k)}(t) = \int \widehat{F}_{C2}^{(k-1)}(t - u) d\widehat{F}_{C2}^{(1)}(u). \qquad (2.5)$$

Although $\widehat{F}_{C2}^{(k)}(t)$ is a biased estimate of $F^{(k)}(t)$ (for $k \geq 2$), it is the nonparametric maximum likelihood estimator. It can also be expressed as

$$\widehat{F}_{C2}^{(k)}(t) = \frac{1}{n^k} \sum_{i_1, i_2, \cdots, i_k} I(T_{i_1} + T_{i_2} + \cdots + T_{i_k} \leq t),$$

$\widehat{F}_{C2}^{(k)}(t)$ is a V-statistic and is closely related to $\widehat{F}_{C1}^{(k)}(t)$, in fact, under some conditions for F it is possible to show that $\widehat{M}_{C2}(t)$, the estimator of $M(t)$ based on (2.4) and (2.5), is also consistent and has the same asymptotic distribution as $\widehat{M}_{C1}(t)$.

The computation of this estimator is also tackled by Schneider et al. [39]. They designate $\widehat{M}_{C2}(t)$ *the empirical renewal function*. Although this estimator is not easy to compute, these authors propose solving the following renewal equation

$$\widehat{M}_{C2}(t) = \widehat{F}_{C2}^{(1)}(t) + \int \widehat{M}_{C2}(t - u)d\widehat{F}_{C2}^{(1)}(u). \qquad (2.6)$$

They propose an efficient method that consists of solving a discretized version of Eq. (2.6), given by

$$\widehat{M}_{C2}^d(r) = \widehat{F}_{C2,d}^{(1)}(r) + \sum_{j=1}^{r} \widehat{M}_{C2}^d(r - j)\left(\widehat{F}_{C2,d}^{(1)}(j) - \widehat{F}_{C2,d}^{(1)}(j - 1)\right).$$

This method involves approximating the empirical distribution by a lattice distribution. The statistical properties of the estimator in (2.6), i.e. consistency and asymptotic normality, are discussed in Grübel and Pitts [19].

More recently, From and Li [15] construct, among other things, nonparametric confidence intervals for $F^{(k)}(t)$ based on the estimator $\widehat{F}_{C2}^{(k)}(t)$. First of all, they give a numerical procedure for approximating the k-fold convolution of F starting with the empirical distribution function. Next, they obtain the asymptotic distribution of $\sqrt{n}\left[\widehat{F}_{C2}^{(k)}(t) - F^{(k)}(t)\right]$ as a Normal law with mean 0 and derive an estimator of the variance. Finally, they give the expression of an approximate $100(1 - \alpha)\%$ confidence interval for $F^{(k)}(t)$. However, as the authors admit, the computational burden is again very high.

2.2.2.2 The Histogram-Type Estimator

Markovich [29] investigates a histogram-type estimator of the renewal function similar to the first Frees estimator. This estimator is based on a new estimator of the k-fold convolution function, where, in contrast to the Frees estimators, only one combination of adjacent renewal times T_i is used.

To describe the estimator, let $[r]$ be the integer part of a real number r. Let $S_k = T_1 + T_2 + \cdots + T_k$, for $k = 1, 2, \ldots$, the waiting times, as defined previously. The estimation of $P\{S_k < t\}$, i.e. the k-fold convolution function of F, is obtained as an empirical distribution function based on an artificially constructed sample of the random variable S_k, from the initial data set of renewal times. For example, to estimate $P\{S_2 < t\} = F^{(2)}(t)$, proceed as follows. Construct the values $\tau_2^i = \sum_{q=2i-1}^{2i} T_q$, for $i = 1, 2, \ldots, n_2 = \left[\frac{n}{2}\right]$.

This procedure produces a sample of size n_2 of the random variable S_2. The associate empirical distribution function is then obtained as

$$\widehat{F}_{HT}^{(2)} = \frac{1}{n_2} \sum_{i=1}^{n_2} I(\tau_2^i \leq t).$$

In a similar way, continue with $k \geq 3$. Define the sequence

$$\tau_k^i = \sum_{q=1+k(i-1)}^{ki} T_q, \text{ for } i = 1, 2, \ldots, n_k = \left[\frac{n}{k}\right].$$

and construct

$$\widehat{F}_{HT}^{(k)} = \frac{1}{n_k} \sum_{i=1}^{n_k} I(\tau_k^i \leq t). \tag{2.7}$$

Note that $n_k = 1$ for $k > n/2$; thus, the estimator above is defined for $k \leq n/2$. The estimator of the renewal function based on (2.7) is therefore given by

$$\widehat{M}_{HT}(k, \kappa) = \sum_{k=1}^{\kappa} \widehat{F}_{HT}^{(k)}(t). \tag{2.8}$$

In the notation, the dependence of the number of terms in the summation is highlighted. The convergence properties of the estimator in (2.8) are investigated in Markovich [29]. The method is based on exploring the error term $\sum_{k=\kappa+1}^{\infty} F^{(k)}(t)$. To do this, some information about F, is required, such as the existence of a moment generating function. The number of terms κ in (2.8) can be determined by two alternative methods. One is to obtain κ, as a function of the sample size n, in order to provide the *a.s.* uniform convergence of the estimator to the true renewal function for small t. The results are established for both light- and heavy-tailed renewal time distributions. The other is to use a plot method to determine a desirable value for κ. The histogram-type estimator is plotted versus κ for fixed t. Then, based on the uniform convergence of the $\widehat{M}_{HT}(t, \kappa)$ to $M(t)$, κ is selected according to

$$\kappa^* = \arg\min \left\{ \kappa : \widehat{M}_{HT}(t, \kappa) = \widehat{M}_{HT}(t, \kappa + 1); \kappa = 1, 2, \ldots, n - 1 \right\}.$$

Compared to Frees estimate, the histogram-type method gives a more computationally tractable estimator. Moreover, Markovich [29] shows that although $\widehat{M}_{HT}(t, \kappa)$ has a greater bias than $\widehat{M}_{C1}(t)$, the mean squared error is smaller. Markovich and Krieger [30] present an alternative data-dependent selection of κ based on a bootstrap method similar to the one we will develop in a subsequent section.

2.2.2.3 Monte-Carlo Estimators

The next group of estimators of the k-fold convolution function of F is based on works by Brown et al. [5] and Gertsbakh and Shpungin [17], who use numerical Monte-Carlo methods to approximate the convolution functions of type $F^{(k)}(t) = P\{T_1 + T_2 + \cdots + T_k \leq t\}$, where T_i are random variables with known CDF, F, for $i = 1, 2, \ldots, k$.

The underlying idea is that the expected value of a random variable may be approximated by generating a large number of samples of the variable and computing the average value toward the samples. Once again, let $T_1, T_2, \ldots, T_n$ be a realization of a renewal process $N(t)$, so that we have a sequence of non-negative independent random variables with CDF F, unknown. Let $F^{(k)}$ be the k-fold convolution function of F. Starting with a random sample from distribution F, our objective is to adapt the k-fold convolutions approximated by the authors cited above. The function F_i's that they use to define their respective procedures are replaced by an empirical distribution based on the sample information. Or, equivalently, the role of each random variable T_i, with known distribution F, is developed by a random variable τ_i with distribution function $\widehat{F}$, for $i = 1, 2, \ldots, k$.

- *The Crude Monte-Carlo estimator*, $\widehat{F}^{(k)}_{CMC}$. The first estimator is easy to implement and is obtained according to the following steps:

  - Simulate N random samples of size k, from the distribution $\widehat{F}$, i.e. the empirical distribution function. Let $t^j_1, t^j_2, \ldots, t^j_k$ be a realization of the jth sample, for $j = 1, 2, \ldots, N$;
  - Define $\varphi^{(j)}(t) = I(t^j_1 + t^j_2 + \ldots + t^j_k \leq t)$, for $j = 1, 2, \ldots, N$;
  - Define $\widehat{F}^{(k)}_{CMC}(t) = \frac{1}{N}\sum_{j=1}^{N} \varphi^{(j)}(t)$.

- *The Brown estimator*, $\widehat{F}^{(k)}_B$. Next, the approximation given by Brown et al. [5] is adapted to the present case. It is obtained according to the following.

  - Define the random variable

$$
Z_k(t) = \begin{cases} \widehat{F}\left(t - \sum_{i=1}^{k-1} \tau_i\right), & \tau_1 + \tau_2 + \cdots + \tau_{k-1} \leq t \\ 0, & \text{otherwise} \end{cases} \tag{2.9}
$$

  where τ_j are considered as independent random variables with distribution function $\widehat{F}$.

  - It is easy to prove that $E[Z_k(t)] = \widehat{F}^{(k)}_{C2}(t)$, the estimator of the k-fold convolution function given in (2.5), where expectation is with respect to $\widehat{F}$.
  - Generate N independent random variables as $Z_k(t)$ defined by (2.9). Denote the jth replication by Z^j_k and approximate the value of the k-fold convolution by

$$
\widehat{F}^{(k)}_B(t) = \frac{1}{N}\sum_{j=1}^{N} Z^j_k(t).
$$

2.2.2.4 Kernel Estimator and Bandwidth Parameter Estimation

Let us now propose a family of smooth estimators for the successive k-fold convolution functions generated from a distribution function F, based on kernel-type estimators. First, given $T_1, T_2, ..., T_n$, a random sample of $i.i.d.$ with CDF F, define an estimator of F by means of

$$\widehat{F}(t,h) = \frac{1}{n}\sum_{i=1}^{n} W\left(\frac{t - T_i}{h}\right),$$

where $W(x) = \int_{-\infty}^{x} w(u)\, du$, [32] with w a kernel function in the context of nonparametric estimation, usually taken to be a non-negative, symmetric function that integrates to one, and h is a bandwidth parameter that controls the amount of smoothness (also called *smoothing parameter*). For our own particular convenience (see [16]), we will consider

$$\widehat{F}_S(t,h) = \frac{1}{n}\sum_{i=1}^{n} \Phi\left(\frac{t - T_i}{h}\right), \tag{2.10}$$

where $\Phi(u)$ is the Gaussian kernel, that is $\Phi\left(\frac{t-T_i}{h}\right)$ represents, for each $i = 1, 2, ..., n$, the distribution function of a Normal law with mean T_i and standard deviation h. In order to estimate the k-fold convolution function of F, we can consider the following estimator

$$\widehat{F}_{S1}^{(k)}(t,h) = \frac{1}{n^k}\sum_{i_1=1}^{n}\cdots\sum_{i_k=1}^{n} \Phi\left(\frac{t - T_{i_1}}{h}\right) * \overset{(k)}{\cdots} * \Phi\left(\frac{t - T_{i_k}}{h}\right). \tag{2.11}$$

The convolution of the kernel functions in the expression (2.11) may be seen as the distribution function of the sum of k independent Normal random variables with standard deviation h, and means $T_{i_1}, T_{i_2}, ..., T_{i_k}$, respectively. With the properties of the Normal family, this is the distribution function of a Normal variable with mean $T_{i_1} + T_{i_2} + \cdots + T_{i_k}$ and standard deviation $\sqrt{k}h$; therefore, it can be noted that

$$\widehat{F}_{S1}^{(k)}(t,h) = \frac{1}{n^k}\sum_{i_1,i_2,...,i_k=1}^{n} \Phi\left(\frac{t - (T_{i_1} + T_{i_2} + \cdots + T_{i_k})}{\sqrt{k}h}\right). \tag{2.12}$$

On the other hand, for large k, the function in (2.12) is tractable only with difficulty from a computational point of view and so we consider a more feasible expression given by

$$\widehat{F}_{S2}^{(k)}(t,h) = \frac{1}{\binom{n}{k}}\sum_{(n,k)} \Phi\left(\frac{t - (T_{i_1} + T_{i_2} + \cdots + T_{i_k})}{\sqrt{k}h}\right), \tag{2.13}$$

where $\sum_{(n,k)}$ denotes the sum over all $\binom{n}{k}$ distinct subsets of size k, $\{i_1, i_2, \ldots, i_k\}$ of $\{1, 2, \ldots, n\}$.

The kernel smoothing of the convolution functions requires the choice of a bandwidth parameter. The general criterion for choosing a value for the smoothing parameter, h, is to minimize some measure of the error of the kernel estimator. One of the most popular measures of such error is the *Mean Integrated Squared Error (MISE)*, defined by

$$MISE(k, h) = E\left[\int \left[F^{(k)}(t) - \widehat{F}_{S2}^{(k)}(t, h) \right]^2 dt \right]. \tag{2.14}$$

So, in principle, expression (2.14) suggests that the choice of h seems to depend on k. However, looking at the different k-fold convolution estimators, the parameter h has been inherited in $\widehat{F}_{S2}^{(k)}(t, h)$ from the smooth estimator in (2.10). So the definitive factor for selecting the bandwidth is to asymptotically minimize the following:

$$MISE(h) = E\left[\int \left[F(t) - \widehat{F}_S(t, h) \right]^2 dt \right], \tag{2.15}$$

which reduces the problem to selecting the bandwidth for a smooth distribution function estimation.

We follow the guidelines given in Hansen [21]. A manageable expression for the asymptotic *MISE* (i.e. *AMISE*) may be obtained using Gaussian kernels. If $h\sqrt{n} \to \infty$ as $n \to \infty$

$$AMISE = \frac{V}{n} - \frac{h}{n\sqrt{\pi}} + \frac{h^4 R_1}{4} + O(h^4), \tag{2.16}$$

which is the result obtained by Jones [23] for the particular case of Gaussian kernels.

The first term to appear in (2.16) does not depend on h, $V = \int_0^\infty F(u)(1 - F(u))du$. Further, $R_1 = \int_0^\infty (d^2 F(u))^2 du$ is a measure of the roughness of F, where d^2 denotes the second derivative operator. This expression can be generalized to

$$R_m = \int_0^\infty \left(d^{m+1} F(u) \right)^2 du,$$

d^{m+1} being the operator indicating the derivative of order $m + 1$, for $m \geq 1$. When using Gaussian kernels, the *AMISE* is minimized for the value of h given by

$$h_0 = \left(\frac{1}{n\sqrt{\pi R_1}}\right)^{\frac{1}{3}}. \tag{2.17}$$

Obviously, the h_0 in expression (2.17) is not known since it depends on the value of R_1 which, in turn, depends on the second derivative of F. Therefore, a plug-in method is used to replace R_1 in (2.17) with a consistent estimate [21].

If F is a Normal distribution with standard deviation σ, $R_1 = (\sigma^3 4\sqrt{\pi})^{-1}$, see Hansen [21], and thus $\widehat{h}_{0,r} = \widehat{\sigma}(4n^{-1})^{1/3}$. This particular estimate of h is called the *reference bandwidth* in Hansen [21]. It will be used later.

According to the plug-in rule developed by Hansen [21], (see Eq. (2.7) therein), it is possible to define the estimator of R_m, for $m \geq 1$, obtained by Jones and Sheather [24], as

$$\widehat{R}_m(b) = (-1)^m \frac{1}{n^2} \sum_{i,j=1}^{n} d^{2m} \phi_b(T_i - T_j), \tag{2.18}$$

where $\phi_b(T_i - T_j)$ is the *pdf* of a Normal variable with mean T_j and standard deviation b, that is, it is a Gaussian kernel with bandwidth given by b. Jones and Sheather [24] show that the optimal b, the one that minimizes the corresponding *AMISE*, depends on R_{m+1} by means of

$$b_m(R_{m+1}) = \left(\frac{2^{m+\frac{1}{2}}\Gamma\left(m+\frac{1}{2}\right)}{\pi n R_{m+1}}\right)^{\frac{1}{2m+3}}.$$

This equation indicates that the b_1 needed to estimate a value of R_1, required for the estimation of h_0 in (2.17), depends on R_2, which must also be estimated. For estimating R_2, a new bandwidth b_2 will be involved that will depend on R_3, and so on. In other words, it could be expressed as

$$\widehat{R}_1 = \widehat{R}_1(R_2) = \widehat{R}_1\left(\widehat{R}_2(R_3)\right) = \cdots = \widehat{R}_1\left(\widehat{R}_2\left(\widehat{R}_3\left(\ldots \widehat{R}_{m-1}(R_m)\right)\right)\right).$$

This relationship suggests the sequential plug-in rule proposed by Hansen [21], which we detail below,

- Fix $N \geq 1$ and take $\widehat{R}_{N+1} = R_{N+1}\left(\widehat{h}_{o,r}\right)$, by means of (2.18), with $\widehat{h}_{o,r}$ the reference bandwidth;
- Obtain recursively, $\widehat{R}_{m-1} = \widehat{R}_{m-1}\left(\widehat{R}_m\right)$, for $m = 2, \ldots, N$;
- Finally, the estimated bandwidth $\widehat{h}_{o,N}$, will result from substituting $\widehat{R}_1 = \widehat{R}_1\left(\widehat{R}_{N+1}\right)$, obtained in the previous step, in Eq. (2.17).

2.3 Alternating Renewal Process

Let us now consider that the renewal procedure is not an instantaneous event, in such a way that the time to repair or replacement cannot be considered negligible. In other words, we now think of a renewal cycle as a two-phase phenomenon, whose duration is determined by two random variables, say, failure time plus renewal time.

2.3.1 Introduction and Some Applications of the ARP in Reliability

A single unit that evolves in time is considered. Only two states are observed for the system: operative and failed. Let $\varphi(t)$ be, by the value zero versus one, the state of the system at time t; thus,

$$\varphi(t) = \begin{cases} 0, & \text{if the system is operative at time } t \\ 1, & \text{otherwise} \end{cases}$$

Let T be the failure time and R the repair time, respectively. It is assumed that the starting state of the system is operative. Many electrical devices respond to this kind of functioning, for example light bulbs simply function or do not function. T and R are completely unknown in the sense that we do not assume any functional form for their distribution functions. In addition, we suppose once again that perfect repairs are carried out on the system, that is, once the system has failed and a repair has been completed, its behavior is exactly the same as if it were new. Under these conditions, $\{\varphi(t), t \geq 0\}$ is an *Alternating Renewal Process (ARP)*.

Let F (f) and G (g) be the cumulative distribution (density) function corresponding to the failure time T and repair time R, respectively, both of which are supposed to be absolutely continuous. We do not assume any parametric distribution family for T and R.

A renewal cycle duration is given by $T + R$. Let $H = F * G$, the *CDF* of $T + R$, where $*$ denotes Stieltjes convolution product. The renewal function is now obtained as $M(t) = \sum_{k=1}^{\infty} H^{(k)}(t)$, where $H^{(k)}(t) = \left(H * \overset{(k)}{\ldots} * H \right)(t)$ is the k-fold convolution. In this case, $H^{(k)}(t) = P\{$"k renewal cycles are completed in $(0, t]$"$\} = P\{(T_1 + R_1) + (T_2 + R_2) + \cdots + (T_k + R_k) \leq t\}$.

Alternating renewal processes have proved their usefulness as stochastic models in many reliability applications. In fact, they have been widely used as models for diverse phenomena in the engineering field. A typical example is the analysis of a machine which periodically fails, undergoes a technical service, which consists of replacement or perfect repair, and is put to work again. This time, non-negligible repair or replacement times are taken into account. An important application is described by Dickey [10], which includes an example that occurs frequently in nuclear safety systems, where a component is continuously monitored with attention to pressure conditions. When the failure is detected, the component is repaired. This situation may be analyzed by using an alternating renewal process.

Another illustrative example where this type of stochastic process appears particularly suitable for modeling is in air-conditioning loads on electrical power systems, as provided by Mortensen [31].

Di Crescenso [9] gives a generalization of the telegrapher's random process, a stochastic process that describes a motion on the real line characterized by two alternating velocities with opposite directions, where the random times separating consecutive reversals of direction perform an *ARP*. The telegrapher's random process has wide applications in diverse areas such as physics, for describing fluorescence intermittency, for example, or in finance, for describing stock prices.

Chen and Yuan [7] calculate performance measures, such as expected value and variance of the transient throughput and the probability that measures the delivery in time for a balanced serial production with no interstage buffers. The work is based on two fundamental assumptions: that each machine alternates between normal and failed, and that up times and down times are i.i.d.; therefore, an *ARP* is considered.

Bernardara et al. [4], present a new model of rain in time. The alternation of meteorological states (namely, wet and dry) is represented by a strict *ARP* with a Generalized Pareto law of wet and dry periods.

Vanderperre and Makhanov [40] introduce a robot safety device system consisting of a robot with internal safety device. The goal is to obtain the availability measures of the system. The system is characterized by the following safety shutdown rule: "Any repair of the failed safety device requires a shut-down of the operative robot". On the other hand, the safety unit must not operate if the robot is under repair. The system is attended to by two different repair men, and any repair is supposed to be perfect and general. The safety device has a constant failure rate and a general repair time. Both the lifetime and the repair time of the robot are general.

The goal in this section is to obtain a nonparametric estimator for the performance measures of a repairable system modeled by a general *ARP*. In particular, we are interested in estimating the point availability and the long-run availability.

2.3.2 Availability Measures of a Repairable System

Availability is probably the most usual measure for the effectiveness of a repairable system. It was defined by Barlow and Proschan [3] as "the probability that the system is operating at a specified time t", which means that the system has not failed in the interval $(0, t]$ or it has been restored after failure so that it is operational at time t. This measure does not tell us how many times the system has failed before t, the availability of a system just quantifies the chance of finding the system operative when it is required. So, availability measures concern both reliability and maintainability properties of the system and increase with improving either time to failure or maintenance conditions.

Different types of availability measures can be established according to underlying criterions, such as time interval considered and the relevant types of maintenance policies. Next, we present different coefficients of availability for single or one-unit system.

- **Instantaneous or Point Availability**, $A(t)$

  Instantaneous availability is the probability that the system will be operational at a given time, t, that is

  $$A(t) = P[\varphi(t) = 0].$$

  When renewals or repairs are not being carried out in the system, the point availability reduces to the reliability function, $A(t) = P\{T > t\}$.

  In case of repairable systems, availability incorporates maintainability information, and therefore, the operative state of a system at an arbitrary time t is guaranteed if either the system has not failed until t or it has successively failed and been repaired and it is functioning properly since the last repair which occurred at time u, $0 < u < t$. As a consequence, it is easy to see that

  $$A(t) = P\{T_1 > t\} + \sum_{k=1}^{\infty} \int_0^t dP\left\{\sum_{i=1}^{k}(T_i + R_i) \le u\right\} P\{T_{k+1} > t - u\}$$

  $$= 1 - F(t) + \int_0^t \sum_{k=1}^{\infty} dH^{(k)}(u)(1 - F(t - u))$$

  $$= 1 - F(t) + \int_0^t (1 - F(t - u))dM(u)$$

  $$= 1 - F(t) + M(t) * (1 - F(t)).$$

  We will return to this expression in Sect. 2.3.5.

- **Average Availability**, $A_{av}(t)$

  This measure gives the proportion of time that the system is available for use. It is calculated as the average value of the point availability function over a period $(0, t]$,

  $$A_{av}(t) = \frac{1}{t} \int_0^t A(u)du,$$

  which may be interpreted as the average proportion of working time of the system over the first t time units in which the system is operative.

- *Steady State Availability*, A

  The steady state or limiting availability is the most commonly used availability measure. It gives the long-run performance of a repairable system and is defined as the limit of the instantaneous availability function as time approaches infinity, that is

  $$A = \lim_{t \to \infty} A(t).$$

As a consequence of the key renewal theorem, an important and useful expression for A can be derived. Classical renewal theory (see [37]) establishes that since $1 - F$ is a bounded function, and, as reasoned previously, it is verified that $A(t) = 1 - F(t) + M(t)*(1 - F(t))$, then, point availability is the unique solution of the equation $A(t) = 1 - F(t) + H(t) *A(t)$ that is bounded on finite intervals. So, by the key renewal theorem, it is deduced, under mild conditions over H, that

$$\lim_{t \to \infty} A(t) = \frac{1}{E[T + R]} \int_0^\infty (1 - F(t)) dt = \frac{E[T]}{E[T] + E[R]}.$$

In other words, it is derived the expression so celebrated in reliability literature that states that

$$A = \frac{MTTF}{MTTF + MTTR},$$

where *MTTF* (*MTTR*) denotes *mean time to failure (repair)*.

In practical applications, it is acceptable that point availability approaches its limiting value after a time period. Thus, it can be thought that after a reasonable period of time the system availability is almost invariant with time. However, in many practical cases, the interest is not in a so long period of time in which a steady situation may have been reached. Consider, for instance, that the useful life of any electrical device, from a user viewpoint, could be much shorter than the time the system availability reaches such a steady value.

Other definitions for the availability of a repairable system could be introduced if we distinguish between different types of maintenance strategies, more explicitly, if we consider only corrective downtime (*inherent availability*) or if shutdowns are scheduled for preventive maintenance (*achieved availability*). The most complex case is when all experienced sources of downtime are considered, such as administrative downtime, logistic downtime, preventive and corrective maintenance downtime. The ratio of the system uptime and total time is then defined as the *operational availability*, and it is the more realistic availability measure in the sense that it is the one that the customer actually experiences. For more details, see for example Kumar et al. [27].

2.3.3 Nonparametric Estimation of Steady-State Availability

The problem of estimating the availability measures of a repairable system has been extensively discussed in the recent literature. Many authors have dealt with this topic in various situations, e.g. Ananda [1] constructs confidence intervals and performs hypotheses testing for the long-run availability of a parallel system with multiple components that have exponential failure and repair times.

Phan-Gia and Turkkan [33] consider a gamma alternating renewal system and obtain several results with regard to the availability function. Although they do not carry out an estimation study of point availability, they do obtain interesting results on the variable representing the random proportion of time that the system is on during a renewal period.

Claasen et al. [8] consider a two unit standby system where random variables involving time duration, i.e. lifetime, repair time, and warm up time for the repair facility, are considered as exponential laws. They obtain an estimator of the steady-state availability under such conditions.

Finally, Hwan Cha et al. [22] and Ke & Chu [25] conduct some procedures for obtaining confidence intervals for the steady-state availability of a repairable system.

The long-run performance of a repairable system is assessed in terms of steady-state availability, which was defined in the previous section as

$$A = \lim_{t \to \infty} P\{\varphi(t) = 0\},$$

the probability that the system is functioning at a large time t. In ARP, it is well known that

$$A = \frac{MTTF}{MTTF + MTTR},$$

where, as defined previously, $MTTF = E[T]$ and $MTTR = E[R]$.

The aim of this section is to conduct inferences on A based on distribution-free estimators of failure and repair time. Let us consider a system that is activated and functioning at time $t = 0$, and replaced by a new one whenever it fails. We observe such a system in a fixed time interval $[0, \tau]$ and let (T_1, R_1), (T_2, R_2), ..., (T_n, R_n) be the registered sample, where $T_1, T_2, ..., T_n$ are the observed lifetimes of the system, which are *i.i.d.* with *CDF F*, $E[T_i] = \mu_T > 0$ and $Var(T_i) = \sigma_T^2$. Likewise, $R_1, R_2, ..., R_n$ are the observed repair times, which are *i.i.d.* with *CDF G*, $E[R_i] = \mu_R > 0$ and $Var(R_i) = \sigma_R^2$.

The natural estimator of A is given by

$$\widehat{A} = \frac{\overline{T}}{\overline{T} + \overline{R}},$$

where $\overline{T}$ and $\overline{R}$ represent the sample means of the T's and R's, respectively.

Let us now derive the asymptotic distribution of $\widehat{A}$, in order to obtain an asymptotic confidence interval for A. To do so, we consider the following function

$$f(x, y) = \frac{x}{x+y},$$

and the Taylor series to first order around the point (a, b), which is given as

$$f(x, y) = f(a, b) + f_x(a, b)(x - a) + f_y(a, b)(y - b),$$

where the subscripts denote the respective partial derivative. Let us consider the last expression for values $x = \overline{T}$; $y = \overline{R}$; $a = \mu_T$ and $b = \mu_R$. As we know,

$$\sqrt{n}(\overline{T} - \mu_T) \xrightarrow{d} N(0, \sigma_T^2)$$

and

$$\sqrt{n}(\overline{R} - \mu_R) \xrightarrow{d} N(0, \sigma_R^2),$$

where $\xrightarrow{d}$ denotes convergence in distribution. Thus, we can write

$$\widehat{A} = \frac{\mu_T}{\mu_T + \mu_R} + \frac{\mu_R}{(\mu_T + \mu_R)^2}(\overline{T} - \mu_T) + \frac{\mu_T}{(\mu_T + \mu_R)^2}(\overline{R} - \mu_R),$$

or equivalently,

$$\sqrt{n}(\widehat{A} - A) = \sqrt{n}\frac{\mu_R}{(\mu_T + \mu_R)^2}(\overline{T} - \mu_T) + \sqrt{n}\frac{\mu_T}{(\mu_T + \mu_R)^2}(\overline{R} - \mu_R).$$

Evaluating the above limits in this expression and using the Delta method, we can obtain

$$\sqrt{n}(\widehat{A} - A) \xrightarrow{d} N(0, \sigma_A^2),$$

where

$$\sigma_A^2 = \frac{\mu_R^2 \sigma_T^2 + \mu_T^2 \sigma_R^2}{(\mu_T + \mu_R)^4}.$$

Let $\widehat{\sigma}_T^2 = \frac{1}{n}\sum_{i=1}^{n}(T_i - \overline{T})^2$ and $\widehat{\sigma}_R^2 = \frac{1}{n}\sum_{i=1}^{n}(R_i - \overline{R})^2$, these being the estimators of σ_T^2 and σ_R^2, respectively. Then we can estimate the variance of the limiting availability by means of

$$\widehat{\sigma}_A^2 = \frac{\overline{R}^2 \widehat{\sigma}_T^2 + \overline{T}^2 \widehat{\sigma}_R^2}{(\overline{T} + \overline{R})^4}.$$

It can be seen that $\widehat{\sigma}_A^2 \xrightarrow{a.s.} \sigma_A^2$, as $n \to \infty$. For a given confidence level $1 - \alpha$, an approximate large sample $100(1 - \alpha)\%$ confidence interval for A can be given by

$$\left(\widehat{A} - z_{\frac{\alpha}{2}} \frac{\widehat{\sigma}_A^2}{\sqrt{n}}, \widehat{A} + z_{\frac{\alpha}{2}} \frac{\widehat{\sigma}_A^2}{\sqrt{n}} \right),$$

with $z_{\frac{\alpha}{2}}$ being the quantile of order $100(1 - \frac{\alpha}{2})\%$ of $N(0,1)$.

2.3.4 Smooth Estimation in the ARP

2.3.4.1 Kernel Estimation

Suppose that the system has been observed up to the nth cycle of the alternating renewal process. During the observation period, we have recorded $0 = S_0 < S_1 < S_2 < \cdots < S_r$, the sequence representing the successive arrival times. For the sake of simplicity in our exposition, we assume that the initial state of the system is operative, and also that the last event recorded is a repair of the system. Under these assumptions, we have no loss of generality, and there exists n such that $r = 2n$.

Let us define the following alternative and independent sequences:

$$T_j = S_{2j-1} - S_{2j-2}, j = 1, 2, \ldots, n,$$

that is, the failure times sequence, and

$$R_j = S_{2j} - S_{2j-1}, j = 1, 2, \ldots, n,$$

the repair times sequence. In other words, we have on the one hand, T_1, T_2, ..., T_n the successive lifetimes of the system, which are *i.i.d.* with *CDF F*. On the other hand, R_1, R_2, ..., R_n, the corresponding repair times, are *i.i.d.* with *CDF G*. We also assume that $\{T_i, R_i\}$ are independent, and therefore we have an *ARP*.

Since the distribution functions F and G are considered to be absolutely continuous, let f and g denote the corresponding density functions. It is possible to give nonparametric estimators of F and G, respectively, based on kernel estimator functions. That is, define

$$\widehat{F}(t, h_1) = \frac{1}{n} \sum_{i=1}^{n} W_1 \left(\frac{t - T_i}{h_1} \right), \tag{2.19}$$

and

$$\widehat{G}(t, h_2) = \frac{1}{n} \sum_{i=1}^{n} W_2 \left(\frac{t - R_i}{h_2} \right), \tag{2.20}$$

where $W_j(x) = \int_{-\infty}^{x} w_j(u)du$, with w_j a kernel function in the context of non-parametric estimation [32], and h_j a bandwidth parameter or smoothing parameter that we need to determine, for $j = 1, 2$.

Remark Given that T and R are non-negative random variables, exponential kernel functions could be used (see Guillamón et al. [20]). That is, $w_1(u) = (1/\bar{t})e^{-u/\bar{t}}$ $I_{[0,+\infty)}(u)$ and $w_2(u) = (1/\bar{r})e^{-u/\bar{r}}I_{[0,+\infty)}(u)$, with $\bar{t}$ and $\bar{r}$ being the observed mean values of failure and repair times, respectively, and $I_{[0,+\infty)}(u) = 1$, for $u \in [0, +\infty)$ and 0, otherwise. One important advantage of using this kind of kernels, which are asymmetric, is that the bias that arises when estimating near the origin is considerably reduced. Nevertheless, we use mainly the Gaussian kernel in our simulation studies.

With respect to the smoothing parameter, we suggest the use of two different values, h_1 and h_2, in the definition of the respective estimators for F and G, since in general, failure and repair times are expected to have different ranges.

2.3.4.2 A Bootstrap Method for Choosing the Bandwidth

One of the most important aspects of kernel estimation is the choice of smoothing parameter. Many different proposals have been made to address this dilemma (see Sect. 2.2.2.4). There exists a vast literature on the use of bootstrap methods for selecting the bandwidth.

Bootstrap resampling techniques were introduced by Efron [11]. One of the earliest references on the subject is the work by Cao [6], who introduced a smooth bootstrap for choosing the bandwidth in kernel density estimation. The method, which exhibited a reasonably reliable behavior, has been subsequently extended to confront the problem of bandwidth selection in other contexts, hazard rate estimation, for instance, and under different sampling schemes, in particular, in the presence of censoring in González-Manteiga et al. [18].

These techniques have already been developed in many reliability applications, producing very good results, see for example, Phillips [34, 35], Marcorin and Abackerli [28] and Gámiz and Román [16]. In all cases, bootstrap techniques reveal significant improvements in estimation compared to traditional techniques. In this section, we present a method based on bootstrap resampling to select the smoothing parameters involved in the kernel estimators of the distribution functions in an *ARP*.

The bandwidth parameter, $\mathbf{h} = (h_1, h_2)'$, can be selected as the minimizer of the mean integrated squared error. In this context of the *ARP*, we define the following *MISE*

$$MISE(\mathbf{h}) = MISE(h_1, F) + MISE(h_2, G), \tag{2.21}$$

where the *MISEs* on the right-hand side are the usual mean integrated squared errors associated with both kernel distribution estimators, this term having been defined previously in Eq. (2.15) as

$$MISE(h_1, F) = E\left[\int \left[\widehat{F}(t, h_1) - F(t)\right]^2 dt\right].$$

This expression is obviously unknown, since we do not have the distribution for which the expectation of (2.21) is calculated. Therefore, we describe below a procedure to approximate the *MISE* in (2.21). This is based on a bootstrap method that consists of imitating the random procedure from which the original sample is drawn, and so we replace the role of the true distribution functions F and G by estimators of the type given by (2.19) and (2.20). In other words, we use the smoothed bootstrap method, where the bootstrap sample is obtained from the estimated values of the distributions F and G. Next, we describe an algorithm for realizing *ARP* trajectories that imitate the original sample. This algorithm is based on the embedded Markov chain. It can be explained as follows:

Let **h** be bandwidth parameters.

Algorithm *Smoothed Bootstrap for ARP*

Step 1. Put $m = 0$, $s_0 = 0$;
Step 2. Generate random variable $T^\bullet \sim \widehat{F}(\cdot; h_1)$ and set $t = T^\bullet(\omega)$;
Step 3. Put $m = m + 1$ and $s_m = s_{m-1} + t$. If $m \geq 2n$ then end;
Step 4. Generate random variable $R^\bullet \sim \widehat{G}(\cdot; h_2)$ and set $r = R^\bullet(\omega)$;
Step 5. Put $m = m + 1$ and $s_m = s_{m-1} + r$. If $m \geq 2n$ then end, otherwise continue to Step 2.

Once the bootstrap sample is drawn, consider the bootstrap version of the estimators in (2.19) and (2.20), $F^\bullet(t, h_1)$ and $G^\bullet(t, h_2)$, for which we have replaced the original sample by the bootstrap sample in the expressions (2.19) and (2.20). Now, define the bootstrap estimate of the mean integrated squared error by

$$MISE^\bullet(\mathbf{h}) = MISE^\bullet(h_1, F) + MISE^\bullet(h_2, G),$$

that is,

$$MISE^\bullet(\mathbf{h}) = E_\bullet\left[\int \left[\widehat{F}^\bullet(t, h_1) - \widehat{F}(t, h_1)\right]^2 dt\right] + E_\bullet\left[\int \left[\widehat{G}^\bullet(t, h_2) - \widehat{G}(t, h_2)\right]^2 dt\right]$$

The minimizer of the above function is the bootstrap bandwidth selector. Although $MISE^\bullet(\mathbf{h})$ can be written in terms of the original sample and, therefore, from a theoretical viewpoint no resampling is needed, an explicit expression is quite hard to obtain (see Cao [6]). Therefore, in practice, Monte-Carlo methods are proposed to calculate the values of $MISE^\bullet(\mathbf{h})$.

The resampling procedure consists in drawing B bootstrap samples in the following way: for each bootstrap sample, b, a replication of $\widehat{F}^\bullet(t, h_1)$ and $\widehat{G}^\bullet(t, h_2)$

are obtained, i.e. $\widehat{F}_b^{\bullet}(t, h_1)$ and $\widehat{G}_b^{\bullet}(t, h_2)$, so that the bootstrap estimation of the standard error may be obtained as the sample mean of the bootstrap samples

$$\widehat{MISE}^{\bullet}(\mathbf{h})$$

$$= \frac{1}{B} \sum_{b=1}^{B} \left[\int \left(\left[\widehat{F}_b^{\bullet}(t, h_1) - \widehat{F}(t, h_1) \right]^2 + \left[\widehat{G}_b^{\bullet}(t, h_2) - \widehat{G}(t, h_2) \right]^2 \right) dt \right]. \quad (2.22)$$

The integral in (2.22) is approximated by numerical methods if necessary. Finally, we choose the vector of bandwidths, $\mathbf{h}_{boot}$ that minimizes the expression (2.22).

In order to perform the resampling procedure, it is necessary to start with a pilot bandwidth $\mathbf{h}^0 = \left(h_1^0, h_2^0 \right)'$ as the initial value and the bootstrap bandwidth parameters are achieved by means of the following iterative method [20]:

$$\widehat{MISE}^{\bullet}(\mathbf{h}^j)$$

$$= \frac{1}{B} \sum_{b=1}^{B} \left[\int \left(\left[\widehat{F}_b^{\bullet}\left(t, h_1^j \right) - \widehat{F}\left(t, h_1^{j-1} \right) \right]^2 + \left[\widehat{G}_b^{\bullet}\left(t, h_2^j \right) - \widehat{G}\left(t, h_2^{j-1} \right) \right]^2 \right) dt \right],$$

where $\widehat{F}_b^{\bullet}\left(t, h_1^j \right)$ and $\widehat{G}_b^{\bullet}\left(t, h_2^j \right)$ are the estimators of the *ARP* distributions based on the bootstrap sample with $\mathbf{h}^j$ as bandwidth vector value; and, $\widehat{F}\left(t, h_1^{j-1} \right)$ and $\widehat{G}\left(t, h_2^{j-1} \right)$ are the estimators based on the original sample and with parameter $\mathbf{h}^{j-1}$, the one that minimizes the $\widehat{MISE}^{\bullet}$ at the previous iteration.

Thus, starting with a pilot bandwidth $\mathbf{h}^0$, the idea is to find the value of $\mathbf{h}^1$ that minimizes $\widehat{MISE}^{\bullet}(\mathbf{h}^j)$, for $j = 1$; then make $j = j + 1$ and repeat the procedure until an appropriate convergence criterion is achieved, that could be, for example that two consecutive $\mathbf{h}^{j-1}$ and $\mathbf{h}^j$ are close enough.

To illustrate this, we apply the method to an alternating renewal process where the failure (repair) times have been simulated from a particular parametric distribution family.

Example Weibull Lifetime and Lognormal Repair Time

First, we consider a system that evolves in time, passing through successive up and down states. The lengths of the up periods are considered to be random variables with Weibull distribution with scale parameter 20 and shape parameter 2. On the other hand, it is known that the Lognormal distribution is a suitable model for the repair times in many cases. We consider a two-parameter Lognormal distribution for the repair time, where the mean log time is chosen as 2 and the standard deviation of the log time is 1.5. Under these conditions, we have simulated 100 renewal cycles, each consisting of an *up* period plus a *down* period. We have applied the smoothed method obtaining the curves given in Fig. 2.1. The bootstrap approximation for the

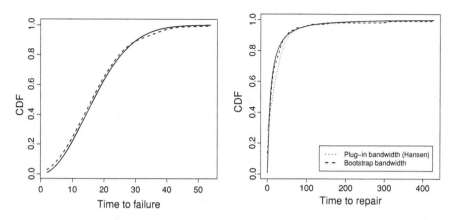

Fig. 2.1 Nonparametric estimation an *ARP*: Theoretical curve (*solid line*) and bootstrap estimate (*dashed line*)

vector of bandwidths is given by $\mathbf{h}_{boot} = (3.159268, 4.543223)'$. These values have been obtained after performing the iterative procedure of the previous section. To do this, we needed a pilot vector of bandwidths to initiate the procedure. We have considered as initial values of the bandwidth parameters the plug-in values suggested in Hansen [21] and given by $\widehat{h}_{0,r} = \widehat{\sigma}(4n^{-1})^{1/3}$, where n is the sample size and $\widehat{\sigma}$ is the sampling standard deviation. In our example, we have obtained as initial vector of bandwidths $\mathbf{h}_{ini} = (3.314656, 18.96642)'$. We stopped when the difference between the values of the $\mathbf{h}'s$ estimated in two consecutives iterations is below 10^{-2}. This convergence was attained after 8 iterations. The results are presented in Fig. 2.1. The solid line in each graph represents the theoretical cumulative distribution function of the Weibull and Lognormal distribution. We have not included the estimator based on the initial value in the left panel, since the difference with the curve obtained with the bootstrap approximation of the bandwidth is not appreciated in the graph. The estimators show a lower precision near the origin. This problem can be solved by means of local estimation procedures or the use of suitable asymmetric kernels; however, this issue will not be addressed here.

2.3.5 Smooth Estimation of the Availability Function

Let $H = F * G$, the *CDF* of a renewal cycle duration, where $*$ denotes Stieltjes convolution product, and $M(t) = \sum_{k=1}^{\infty} H^{(k)}(t)$, where $H^{(k)}(t) = \left(H * \overset{(k)}{\cdots} * H\right)(t)$ is the k-fold convolution.

As shown in classical renewal theory (see Sect. 2.3.2), the availability function satisfies the following renewal type equation,

$$A(t) = 1 - F(t) + \int_0^t (1 - F(t - u))dM(u) = 1 - F(t) + M(t) * (1 - F(t))$$

$$(2.23)$$

which is generally not easy to evaluate.

The estimation of the availability is now based on kernel estimators of the functions F and G defined in expressions (2.19) and (2.20), respectively, hence

$$\widehat{A}(t, \mathbf{h}) = 1 - \widehat{F}(t, h_1) + \widehat{M}(t, \mathbf{h}) * \left(1 - \widehat{F}(t, h_1)\right) \qquad (2.24)$$

where $\widehat{M}(t, \mathbf{h}) = \sum_{k=1}^{\infty} \widehat{H}^{(k)}(t, \mathbf{h})$. Expression (2.24) depends on $\mathbf{h} = (h_1, h_2)'$, the bootstrap bandwidth parameters for estimating F and G, respectively, obtained previously.

The aim is to give the value of the above expression for the availability. For this purpose, note that the most tedious problem to arise in expression (2.23) is that of deriving the renewal function $M(t)$, and therefore we proceed as follows. First, we approximate the value of $M(t)$ by plug-in into the expression of M (given above), the functions F and G by their respective exponential kernel estimator as explained in the previous section, that is (2.19) and (2.20), respectively.

So, for estimating the k-fold convolution function $H^{(k)}(t)$ we have the following expression

$$\widehat{H}_n^{(k)}(t, \mathbf{h}) = \frac{1}{n^{2k}} \sum_{i_1, \dots, i_k; j_1, \dots, j_k} W_1\left(\frac{t - T_{i_k}}{h_1}\right) * \cdots * W_1\left(\frac{t - T_{i_k}}{h_1}\right)$$

$$* W_2\left(\frac{t - R_{j_1}}{h_2}\right) * \cdots * W_2\left(\frac{t - R_{j_k}}{h_2}\right) \qquad (2.25)$$

for $k = 1, 2, \dots$; where $\{T_{i_m}\}$ and $\{R_{j_m}\}$ are the observed failure and repair times; $\mathbf{h}_{boot} = (h_1, h_2)'$ is the bootstrap bandwidth vector and, W_1 and W_2 represent the CDF of an exponential random variable with scale parameters $\bar{t}$ and $\bar{r}$, respectively, i.e. the observed sample means. For each $i, j = 1, 2, \dots, n$, the convolution into the sum can be viewed as the CDF of the sum of two independent random variables, one with exponential distribution with scale parameter $h_1\bar{t}$ and location T_i, and the other with scale $h_2\bar{r}$ and location R_j. So, these convolutions can be obtained analytically (we have used the package $distr$ [38] of R for that purpose).

Given that we have chosen exponential kernel functions, the convolutions that appear in expression (2.25) can be simplified as follows

$$\widehat{H}_n^{(k)}(t, \mathbf{h}) = \frac{1}{n^{2k}} \sum_{i_1, \dots, i_k; j_1, \dots, j_k} \mathbf{W}_1\left(\frac{t - (T_{i_1} + \cdots + T_{i_k})}{h_1}\right)$$

$$* \mathbf{W}_2\left(\frac{t - (R_{j_1} + \cdots + R_{j_k})}{h_2}\right),$$

Table 2.1 Kernel estimation of the renewal function

t_0	1	3	5	6	7	10	15
$\widehat{M}(t_0)$	1.0042	1.5492	2.4998	2.9183	3.3512	4.57678	6.8928
	(1.2869)	(2.2393)	(3.0631)	(3.4552)	(3.8391)	(4.9550)	(6.6132)

where W_1 and W_2 are respectively, the distribution functions of the corresponding Gamma family.

To conclude, the number of terms, k_0, considered in $M(t)$ can be determined by the normal approximation of the number of renewals in $(0, t]$, that is, if $Y_i = T_i + R_i$ is the length of a renewal cycle $k_0 = \min\{k : P[Y_1 + Y_2 + \cdots + Y_k \leq t] \leq \varepsilon\}$, with ε fixed small enough.

Example Kernel Estimation of the Renewal Function

To study the empirical performance, we carried out the following simulation study. Let $Y_1, Y_2, \ldots, Y_n$ be the length of $n = 10$ renewal cycles, each obtained by means of a life period simulated from a *CDF* Weibull with scale $\beta = 1$ and shape $\alpha = 2$, plus a down period simulated from a *CDF* Weibull with scale $\beta = 1$ and shape $\alpha = 0.5$. Next, we applied the bootstrap procedure as indicated above to estimate the distributions associated with the *ARP*.

A computational procedure that gives the value of the renewal function $M(t_0)$ for a fixed t_0 may be implemented by means of some functions working in R. First, the method approximates the number of significant terms in $M(t_0)$, i.e. k_0 (defined above); and then it uses the convolution function of gamma distributions, which is implemented in the package *distr* of R, to obtain the probability distribution functions that appear in the successive $\widehat{H}^{(k)}$. Some of the results are displayed in Table 2.1.

The bootstrap approximation of the bandwidth involved in these calculations is given by $\mathbf{h} = (0.0505, 0.1304)'$. The numbers in parentheses express the values of $M(t_0)$ provided by using the functions that approximate the convolution of absolutely continuous distributions, performed in the package *distr* of R. These values are used here for reference.

In any case, from a computational point of view, expression (2.24) is extremely awkward to evaluate, so in Sect. 2.3.7, we propose a slightly more feasible procedure.

2.3.6 *Consistency of the Bootstrap Approximation of the Availability Function*

The Laplace transform may be used to prove the consistency of the estimate defined in (2.24), which is established in terms of the *unavailability function*, that is, $U(t) = 1 - A(t)$.

Taking the Laplace transform on both sides of Eq. (2.23), with the properties of the Laplace transform, we obtain

$$\widehat{U}^*(s, \mathbf{h}) = \frac{1}{s} - \widehat{A}^*(s, \mathbf{h}) = \frac{1}{s} - \frac{1 - \widehat{f}^*(s, h_1)}{s\left[1 - \widehat{f}^*(s, h_1)\widehat{g}^*(s, h_2)\right]}, \tag{2.26}$$

where $\widehat{f}^*(s, h_1)$ and $\widehat{g}^*(s, h_2)$ represent respectively the Laplace transform of the kernel densities, which are

$$\widehat{f}^*(s, h_1) = w^*(sh_1)\frac{1}{n}\sum_{i=1}^{n} e^{-sT_i}$$

and

$$\widehat{g}^*(s, h_2) = w^*(sh_2)\frac{1}{n}\sum_{i=1}^{n} e^{-sR_i},$$

given that we have used the same kernel function, W, in the estimation of F and G.

In the above expressions, $w^*(sh_j)$ is the Laplace transform of the function $w = dW$, say the derivative of W, evaluated in $sh_j, j = 1, 2$. That is, $w^*(s) = \int_0^\infty e^{-st}w(t)dt$. We assume that the kernel function w is such that its Laplace transform exists for $s > 0$ (which is valid in the case of Gaussian or exponential kernels, for example).

Considering expression (2.26), we find that

$$\widehat{U}^*(s, \mathbf{h}) = \frac{1}{s} - \frac{1 - \widehat{w}^*(sh_1)\frac{1}{n}\sum_{i=1}^{n} e^{-sT_i}}{s\left[1 - \left(\widehat{w}^*(sh_1)\frac{1}{n}\sum_{i=1}^{n} e^{-sT_i}\right)\left(\widehat{w}^*(sh_2)\frac{1}{n}\sum_{i=1}^{n} e^{-sR_i}\right)\right]}. \tag{2.27}$$

If the Laplace transform f^* does exist, we find that $f^*(s) = E[e^{-sT}]$. Since $\{T_i\}$ are i.i.d. with density function f, by the strong law of large numbers, for any s for which the above expectation exists,

$$\frac{1}{n}\sum_{i=1}^{n} e^{-sT_i} \overset{a.s.}{\longrightarrow} f^*(s) \text{ as } n \to \infty,$$

and the same argument is valid for establishing that

$$\frac{1}{n}\sum_{i=1}^{n} e^{-sR_i} \overset{a.s.}{\longrightarrow} g^*(s) \text{ as } n \to \infty.$$

Moreover, for fixed s, since $h_j \to 0$ as $n \to \infty$, with the properties of the Laplace transform, we obtain $w^*(sh_j) \to 1$, for $j = 1, 2$. In conclusion, from (2.27), we obtain

$$\widehat{U}^*(s,\mathbf{h}) \longrightarrow \frac{1}{s} - \frac{1 - f^*(s)}{s[1 - f^*(s)g^*(s)]}.$$

The right-hand side of the last expression corresponds to the Laplace transform of the unavailability function $U(t)$.

The unavailability function may be considered as a defective measure, so that we can apply the *extended continuity theorem of the Laplace transform for measures* (see [12]), according to which, if H_n is a measure with the Laplace transform φ_n, for $n = 1, 2, \ldots,$ and $\varphi_n(s) \to \varphi(s)$ for $s > 0$, then φ is the Laplace transform of a measure H and $H_n \to H$, and this convergence is for each bounded interval of continuity of H.

In conclusion, we deduce the uniformly strong consistency of the estimator of availability given by (2.23), that is, for all $t \in R_+$,

$$\sup_{t \in [0,\tau]} \left| \widehat{A}(t,\mathbf{h}) - A(t) \right| \xrightarrow{a.s.} 0, \text{ as } n \to \infty.$$

The Laplace transforms described above, all correspond to defective distributions since, as can easily be checked, for any $n = 1, 2, \ldots, \lim_{s \to 0} \widehat{U}^*(s,\mathbf{h}) \neq 1$. This result may be established by means of the following limit

$$\lim_{s \to 0} s\widehat{A}^*(s,\mathbf{h}) = \frac{h_1 + \bar{t}_n}{(h_1 + \bar{t}_n) + (h_2 + \bar{r}_n)}, \tag{2.28}$$

which is independent of the kernel function w. Here, $\bar{t}_n$ and $\bar{r}_n$ represent the mean sample values.

By the properties of the Laplace transform, we find that $\lim_{s \to 0} s\widehat{A}^*(s,\mathbf{h}) = \lim_{t \to \infty} \widehat{A}(t,\mathbf{h})$. Taking this property together with (2.28), we obtain via our kernel estimator of the availability, an expression for asymptotic availability which is congruent with the known result that establishes that

$$\lim_{t \to \infty} A(t) = \frac{MTTF}{MTTF + MTTR},$$

where *MTTF* denotes the mean time to failure and *MTTR*, the mean time to repair.

Example: Exploring Exponential Kernel Functions

The estimation procedure above may also be carried out by considering the sample information given by the observed times between failures, i.e. $Y_i = T_i + R_i$, for $i = 1, 2, \ldots, n$. Let $H(t)$ denote the theoretical cumulative distribution function; in this case, kernel estimation of the availability function would have the following Laplace transform

$$\widehat{A}^*(s,\mathbf{h}) = \frac{1 - w^*(s,h_1)\frac{1}{n}\sum_{i=1}^{n} e^{-sT_i}}{s\left[1 - w^*(s,h_2)\frac{1}{n}\sum_{i=1}^{n} e^{-sY_i}\right]}.$$

If we consider an exponential kernel function, the last expression is of the form

$$\widehat{A}^*(s,\mathbf{h}) = \frac{1 - \left(\frac{1}{sh_1+1}\right)\frac{1}{n}\sum_{i=1}^n e^{-sT_i}}{s\left[1 - \left(\frac{1}{sh_2+1}\right)\frac{1}{n}\sum_{i=1}^n e^{-sY_i}\right]}.$$

We can approximate the above exponential functions by their Taylor expansions, obtaining, for s near 0, that

$$\widehat{A}^*(s,\mathbf{h}) = \frac{1 - \left(\frac{1-s\bar{t}_n}{sh_1+1}\right)}{s\left[1 - \left(\frac{1-s\bar{y}_n}{sh_2+1}\right)\right]},$$

where $\bar{t}_n$ is the mean failure time and $\bar{y}_n$ is the mean duration of a renewal cycle, that is $\bar{y}_n = \bar{t}_n + \bar{r}_n$. Easy computations lead to

$$\widehat{A}^*(s,\mathbf{h}) = \frac{h_1 + \bar{t}_n}{h_2 + \bar{y}_n}\left(\frac{sh_2 + 1}{s(sh_1 + 1)}\right),$$

and, by inverting this expression, we obtain an estimator of the availability by the following

$$\widehat{A}(t,\mathbf{h}) = \frac{h_1 + \bar{t}_n}{h_2 + \bar{y}_n}\left[1 + \left(\frac{h_2}{h_1} - 1\right)e^{-\frac{t}{h_1}}\right],$$

which is valid for large values of t, given the equivalence between the values of the Laplace transform of a function, in this case A^*, near the origin, and the asymptotic values of the function, say A.

2.3.7 Bootstrap Estimate of the k-Fold Convolution of a Distribution Function

In Sect. 2.2.2.4, a kernel estimator for the k-fold convolution function was defined. There, a plug-in bandwidth selector was suggested, based on the asymptotic form of the mean integrated squared error (*MISE*).

In this section, we use a procedure based on bootstrap techniques similar to those in Sect. 2.3.4, in order to find the value of h that minimizes the *MISE*. As in Sect. 2.2.2.4, the particular form of $\widehat{F}_{S2}^{(k)}(t,h)$ for all $k \geq 2$ reduces the problem of finding a bandwidth for any k to the first step, that is for $k = 1$. So, the optimization problem is stated as finding the value of h that minimizes

$$MISE(h) = E\left[\int \left[F(t) - \widehat{F}_{S2}(t,h)\right]^2 dt\right].$$

Using a similar rationale to that given in Sect. 2.3.4.2, the bootstrap bandwidth parameter is achieved by means of the following iterative method

$$\widehat{MISE}(h^j) = \frac{1}{B}\sum_{b=1}^{B}\left\{\int\left[\widehat{F}_b^{\bullet}(t,h^j) - \widehat{F_{S2}}(t,h^{j-1})\right]^2 dt\right\}.$$

We carry out a smoothed bootstrap to obtain bootstrap samples in the observation interval, which is determined, as above, by the occurrence of the nth renewal, the size of the original sample.

Algorithm *Smoothed Bootstrap*

Step 1. Put $m = 0$ and $s_0 = 0$;

Step 2. Generate random variable $T^{\bullet} \sim \widehat{F}(\cdot; h)$ and set $t = T^{\bullet}(\omega)$;

Step 3. Put $m = m + 1$ and $s_m = s_{m-1} + t$. If $m \geq n$ then end, otherwise continue to Step 2.

Example: Kernel Estimation of the k-fold Convolution of a Distribution Function

To illustrate, we now consider the following simulation study. Let $T_1, T_2, ..., T_n$ be the lengths of $n = 10$ simulated renewal cycles, i.i.d. with *CDF* Weibull with scale parameter $\beta = 1$ and shape parameter $\alpha = 3$. We construct the estimator $\widehat{F}_{S2}^k(t, h)$, as defined in Eq. (2.13), for $k = 2, 3, 4, 5, 6, 7$, and obtain the corresponding values for t in [0, 10]. The approximation of bandwidth is obtained by bootstrap techniques in the first step, that is for $k = 1$, which gives the value $h_{boot} = 0.2786$. The results are displayed in the figures below. We compared our results to those obtained with the estimator proposed by Frees [14], that is, M_{C1} which is defined in Sect. 2.2.2.1 based on Eq. (2.3), and found that the kernel estimator gives greater accuracy. We made use once again of the *distr* package provided by the R programming system in order to approximate, for $k = 2, 3, 4, 5, 6, 7$, the convolution functions, unfeasible in theory, for the Weibull distribution F with scale parameter $\beta = 1$ and shape parameter $\alpha = 3$. We use the functions performed under R as reference values to contrast the accuracy of the

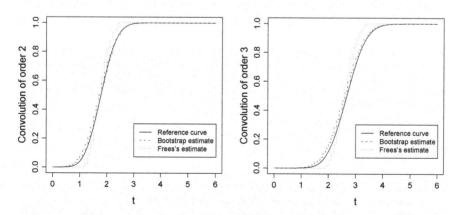

Fig. 2.2 Nonparametric estimation of the k-fold convolution, for $k = 2, 3$

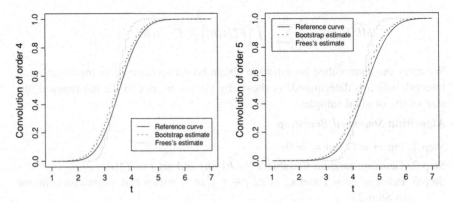

Fig. 2.3 Nonparametric estimation of the k-fold convolution, for $k = 4, 5$

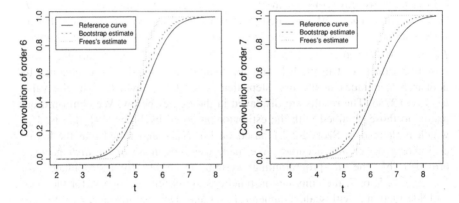

Fig. 2.4 Nonparametric estimation of the k-fold convolution, for $k = 6, 7$

two estimators. The bootstrap curve captures the "theoretical" behavior better than Frees's curve, as can be appreciated from Figs. 2.2, 2.3, 2.4.

Acknowledgments Section 2.3 of this chapter is an extension into book-length form of the article *Nonparametric estimation of the availability in a general repairable system*, originally published in *Reliability Engineering and System Safety* **93** (8), 1188–11962 (2008).The authors express their full acknowledgement of the original publication of the paper in the journal cited above, edited by Elsevier.

References

1. Ananda MMA (1999) Estimation and testing of availability of a parallel system with exponential failure and repair times. J Stat Plan Inf 77:237–246
2. Ascher H, Feingold H (1984) Repairable systems reliability: modelling, inference, misconceptions and their causes. Marcel Dekker Inc, New York

3. Barlow RE, Proschan F (1975) Statistical theory of reliability and life testing. Probability Models. Holt, Rinehart and Winston, New York

4. Bernardara P, De Michele C, Rosso R (2007) A simple model of rain in time: an alternating renewal process of wet and dry states with a fractional (non-Gaussian) rain intensity. Atmos Res 84:291–301

5. Brown M, Solomon H, Stevens MA (1981) Monte-Carlo simulation of the renewal function. J Appl Probab 13:426–434

6. Cao R (1993) Bootstrapping the mean integrated squared error. J Multivar Anal 45:137–160

7. Chen C-T, Yuan J (2003) Throughput analysis using alternating renewal process in balanced serial production lines with no interstage buffers. J Chin Insst Ind Eng 20(5):522–532

8. Claasen SJ, Joubert JW, Yadavalli VSS (2004) Interval estimation of the availability of a two unit standby system with non instantaneous switch over and 'dead time'. Pak J Stat 20(1):115–122

9. Di Crescenso A (2001) On random motions with velocities alternating at Erlang-distributed random times. Adv Appl Probab 33(3):690–701

10. Dickey JM (1991) The renewal function for an alternating renewal process, which has a Weibull failure distribution and a constant repair time. Reliab Eng Syst Saf 31:321–343

11. Efron B (1979) Bootstrap methods: another look at the jackknife. Ann Stat 7:1–26

12. Feller W (1971) An introduction to probability theory and its applications, vol II, 2nd edn. Willey, New York

13. Frees EW (1986a) Warranty analysis and renewal function estimation. Naval Res Logist Quart 33:361–372

14. Frees EW (1986b) Nonparametric renewal function estimation. Ann Stat 14(4):1366–1378

15. From SG, Li L (2003) Nonparametric estimation of some quantities of interest from renewal theory. Naval Res Logist 50(6):638–649

16. Gámiz ML, Román Y (2008) Non-parametric estimation of the availability in a general repairable system. Reliab Eng Syst Saf 93:1188–1196

17. Gerstbakh I, Shpungin Y (2004) Renewal function and interval availability: a numerical Monte Carlo study. Commun Stat Theory Methods 33(3):639–649

18. González-Manteiga W, Cao R, Marron JS (1996) Bootstrap selection of the smoothing parameter in nonparametric hazard rate estimation. J Am Stat Assoc 91(435):1130–1140

19. Grübel R, Pitts SM (1993) Nonparametric estimation in renewal theory I: the empirical renewal function. Ann Stat 21(3):1431–1451

20. Guillamón A, Navarro J, Ruiz JM (1999) A note on kernel estimators for positive valued random variables. Sankhya: Indian J Stat 61 Ser A (Part 2):276–281

21. Hansen BE (2004) Bandwidth selection for nonparametric distribution estimation. Discussion paper, University of Wisconsin, Madison

22. Hwan Cha J, Sangyeol L, Jongwoo J (2006) Sequential confidence interval estimation for system availability. Qual Reliab Eng Int 22:165–176

23. Jones MC (1990) The performance of kernel density functions in kernel distribution function estimation. Stat Probab Lett 9:129–132

24. Jones MC, Sheather SJ (1991) Using non-stochastic terms to advantage in kernel-based estimation of integrated squared density derivatives. Stat Probab Lett 11:511–514

25. Ke JC, Chu YK (2007) Nonparametric analysis on system availability: confidence bound and power function. J Math Stat 3(4):181–187

26. Kijima M (1989) Some results for repairable systems with general repair. J Appl Probab 26(1):89–102

27. Kumar UD, Crocker J, Chitra T, Saranga H (2006) Reliability and six sigma. Springer, New York

28. Marcorin AJ, Abackerli AJ (2006) Field failure data: an alternative proposal for reliability estimation. Qual Reliab Eng Int 22:851–862

29. Markovich NM (2004) Nonparametric renewal function estimation and smoothing by empirical data. Preprint ETH, Zuerich

30. Markovich NM, Krieger UR (2006) Nonparametric estimation of the renewal function by empirical data. Stoch Models 22:175–199
31. Mortensen RE (1990) Alternating renewal process models for electric power system loads. IEEE Trans Autom Control 35(11):1245–1249
32. Mugdadi AR, Ghebregiorgis GS (2005) The Kernel distribution estimator of functions of random variables. Nonparametr Stat 17(7):807–818
33. Pham Gia T, Turkkan N (1999) System availability in a gamma alternating renewal process. Naval Res Logist 46:822–844
34. Phillips MJ (2000) Bootstrap confidence regions for the expected ROCOF of a repairable system. IEEE Trans Reliab 49(2):204–208
35. Phillips MJ (2001) Estimation of the expected ROCOF of a repairable system with bootstrap confidence region. Qual Reliab Eng Int 17:159–162
36. Rausand M, Hoyland A (2004) System reliability theory: models, statistical methods and applications, 2nd edn. Wiley, New York
37. Ross SM (1992) Applied probability models with optimization applications. Courier Dover Publications, New York
38. Ruckdeschel P, Kohl M, Stabla T, Camphausen F (2006) S4 classes for distributions. R News 6(2):2–6. http://www.uni-bayreuth.de/departments/math/org/mathe7/DISTR/distr.pdf
39. Schneider H, Lin B-S, O'Cinneide C (1990) Comparison of nonparametric estimation for the renewal function. Appl Stat 39(1):55–61
40. Vanderperre EJ, Makhanov SS (2008) Point availability of a robot with internal safety device. Contemp Eng Sci 1(1):15–25

Chapter 3
Models for Minimal Repair

3.1 Introduction

3.1.1 The Concept of Minimal Repair

In this chapter, systems subject to failure and repair are considered. Once the failure in the system has occurred, certain maintenance activities are carried out and the system is returned to operational conditions. To formulate a model for the system's evolution, it is essential to describe the characteristics of the system once the maintenance action has been completed. For this reason, in the specialized literature, models for repairable systems have been distinguished according to the effect the failure and consequent repair have on the system's behavior.

In the previous chapter, we considered that once a failure had occurred, its functioning would be restored by either replacing the system with a new one or repairing it in such a way that its operational conditions would be exactly the same as if it were new. The system is then said to be in "as good as new" condition. Perfect repairs are carried out on the system, and so the appropriate model for this case is a renewal process (RP), to which Chap. 2 is devoted.

However, complex systems are usually repaired rather than replaced after failure. In practice, it is more convenient, mainly from an economic viewpoint, to restore the system to functionality by methods other than replacing the entire system. In particular, many repairs in real life return the system to operational conditions that are basically the same as they were just before the failure occurred. As pointed out in Aven [7], the goal is to return the system to a functioning condition as soon as possible and therefore it seems reasonable that, after repair, the system is brought to the exact same state it was in just before the failure. We refer to this situation by saying that the system is in "as bad as old" condition, which leads us to the concept of minimal repair introduced by Ascher [3] where it

M. L. Gámiz et al., *Applied Nonparametric Statistics in Reliability*,
Springer Series in Reliability Engineering, DOI: 10.1007/978-0-85729-118-9_3,
© Springer-Verlag London Limited 2011

is considered that the "age" (in the sense of effectiveness) of the system after repair is the same as it was before failure.

The situation above frequently arises in the maintenance of complex systems (with many components), where after failure, small parts of the system are replaced by new ones. In such a case, perfect repairs are restricted to component-level. If we adopt the Ascher formulation [4], a system is understood to be a repairable item, whereas a component in the system is considered to be non-repairable, and therefore, when a component of the system fails, it is replaced by a new one. With this understanding, if one component of a complex system fails and is replaced, the system as a whole can be considered to be in approximately the same state as that immediately before the failure.

At this point, it is worth distinguishing between the two interpretations of minimal repair traditionally considered, which are the so-called *physical* minimal repair and *statistical* or *black-box* minimal repair. The difference between the two concepts lies in the level of information available about the system itself. In the physical minimal repair, information about the states (working or failed) of the components of the system is available, so the failed component is localized and then immediately replaced. In the case of systems with many components, it can be assumed that the event of failure-replacement of a particular component does not affect the age of the system as a whole. Obviously, the information about components may introduce different ways of understanding the performance level of complex systems (see Chap. 5 about multi-component systems). In the case of statistical minimal repair, only the failure of the whole system is recognized. The description of the state of the system is formulated exclusively in terms of its age and does not contain any additional information. With this interpretation, we understand that the failed system is replaced by a statistically identical one, which was running under the same operating conditions but has not yet failed.

In this chapter we will consider the type of minimal repair represented by the statistical minimal repair, concentrating on *black-box* modeling and analysis. In other words, we do not have information about the design of the system and therefore we ignore any further information about the state of the components, formulating the models for the pattern of system-level failures.

3.1.2 Basic Concepts for Counting Processes

Since we are dealing with repairable systems, we consider that the system suffers multiple failures over time, so we need a stochastic model to describe the occurrence of events (that are failures of the system) in time, i.e. a point process. In general, the elapsed times between successive failures are neither independent nor identically distributed as was the case with models for perfect repair considered in the previous chapter. Let us now establish some of the concepts we will use in this and the next chapter.

Consider a repairable system which is put into operation for the first time at $S_0 = 0$, let $S_1 < S_2 < S_3 < ...$, be the ordered random times at which the successive failures occur. We ignore the time duration of repairs, so we assume that once the system fails, it is immediately put into operation. The inter-arrival times will be denoted as $T_i = S_i - S_{i-1}$, for $i \geq 1$. A stochastic model for the evolution of the system will be required, and it should properly specify the joint distribution of $\{T_1, T_2, ...\}$.

The corresponding counting process is the stochastic process $\{N(t); t \geq 0\}$ where $N(t)$ denotes the number of failures that occur in the interval $(0, t]$. Thus, $N(t_1, t_2) = N(t_2) - N(t_1)$ counts the number of failures to have occurred in the interval $(t_1, t_2]$. It is noticeable that (with probability 1) $N(t)$ is a non-decreasing, right-continuous step function. A basic reference in the treatment of counting processes is Andersen et al. [2]. See also Rausand and Høyland [28] for an introduction to the theory.

The next definitions introduce some key functions for describing the evolution of repairable systems.

Definition 3.1 (*Mean Function of a Counting Process*) Let $\{N(t); t \geq 0\}$ be a counting process, the mean function of the process is defined as the expectation

$$M(t) = E[N(t)], \quad \text{for} \quad t \geq 0. \tag{3.1}$$

Therefore, $M(t)$ is the expected number of failures up to time t and is a non-decreasing function of t. It can also be verified that $M(t)$ is a right-continuous function (see for example [29]). The next definition gives an important measure that characterizes some types of counting processes.

Definition 3.2 (*Rate of Occurrence of Failures*) (*Rocof*) When the mean function $M(t)$ is differentiable, the following function can be defined

$$m(t) = \frac{\mathrm{d}}{\mathrm{d}t} M(t), \quad \text{for} \quad t \geq 0. \tag{3.2}$$

This function is usually interpreted as the instantaneous rate of change in the mean number of failures.

If the system is subject to perfect repairs (as described in Chap. 2) and hence the underlying model is a renewal process, the function $M(t)$ and its first derivative $m(t)$ as given in Definitions 3.1 and 3.2 play a central role and are referred to as *renewal function* and *renewal density*, respectively.

A closely related function is given next.

Definition 3.3 (*Unconditional Intensity Function*) Let $\{N(t); t \geq 0\}$ be a counting process, the (unconditional) intensity function is given as

$$\mu(t) = \lim_{\varepsilon \to 0} \frac{P\{N(t+\varepsilon) - N(t) = 1\}}{\varepsilon}. \tag{3.3}$$

The probability of failure in a small interval can be approximated by the product of the intensity function and the length of the interval. In an interval where $\mu(t)$ is large, many failures are expected, whereas in intervals where $\mu(t)$ is small, few failures are expected.

When the probability of simultaneous failures is zero, that is when the process is *orderly* (as is the case of the processes we will consider in the following chapters), it can be shown (see [29]) that the intensity function equals the rocof, that is $\mu(t) = m(t)$.

It is worth noting that, contrary to the hazard function of a random variable, i.e. $\lambda(t)$, the intensity function is not defined in terms of a conditional probability. In other words, the intensity function is the unconditional probability of a failure at a small interval divided by the length of the interval. Clearly, this is a measure associated with repairable systems while the hazard function is not. In the definition of the intensity function, nothing is said about the history of the system until t, the only interest is in the occurrence of the next failure, which is not necessarily the first. In the definition of the hazard function, the history of the (non-failed) system until the present moment appears implicit, since the probability of failure (now the first) given that the system has not yet failed is being calculated.

While the mean function $M(t)$, the rocof $m(t)$ and the unconditional intensity $\mu(t)$ are useful concepts in the description of a counting process, they do not completely characterize the stochastic properties of the process. Instead, a more complete information index is necessary, given in the following definition.

Definition 3.4 (*Conditional Intensity Function*) Let $\mathcal{F}_{t^-}$ denote the history of the process up to, but not including, time t. $\mathcal{F}_{t^-}$ is generated by the set $\{N(s), 0 \le s < t\}$, and so it contains all the information about failure times in the past, that is until t, and is called a *filtration*. The conditional intensity function is defined as

$$\gamma(t) = \lim_{\varepsilon \to 0} \frac{P\{N(t+\epsilon) - N(t) \ge 1 \mid \mathcal{F}_{t^-}\}}{\varepsilon} \tag{3.4}$$

This function should not, in general, be confused with the rocof $m(t)$ given in Definition 3.2, nor, of course, with the unconditional intensity function $\mu(t)$. However, in the particular case of NHPP, as we shall see below, the conditional intensity depends on the history of the process only through time t, that is, unlike other more complex models (such as Cox processes, renewal processes, or multiplicative intensity models, among others), for a NHPP, $\gamma(t)$ is a (deterministic) function of only time t and it is true that $\gamma(t) = m(t)$.

3.1.3 The Model: Non-Homogeneous Poisson Process (NHPP)

Consider a repairable system, with the notation as in the previous section. Let us denote F as the cumulative distribution function corresponding to $T = T_1$ (time until the system fails for the first time); $\bar{F} = 1 - F$ the survival or reliability function and $\lambda(t)$ the hazard function, then

$$\bar{F}(t) = \exp\left[-\int_0^t \lambda(u)du\right]. \tag{3.5}$$

Next, we give the mathematical definition of minimal repair, which was introduced by Ascher [3]. See Hollander et al. [18] or Finkelstein [15] for recent references.

Definition 3.5 (*Minimal Repair*) The survival function of a system that has failed and been *instantaneously minimally repaired* at age s is defined as

$$\bar{F}_s(t) = \frac{1 - F(s+t)}{1 - F(s)} = \exp\left[-\int_s^{s+t} \lambda(u)du\right]. \tag{3.6}$$

From the definition above it can be deduced that, given that the observed time of the first failure is $S_1 = s_1$ the survival function corresponding to the waiting time to the second failure under minimal repair is given by $\bar{F}_{s_1}$. In general, when minimal repairs are being performed in the system, given that the jth failure has occured at time $S_j = s_j$, we have that

$$P\{T_{j+1} > t | S_j = s_j\} = \bar{F}_{s_j}(t), \quad \text{for} \quad t \geq 0, \quad j = 1, 2, \ldots \tag{3.7}$$

is the survival function corresponding to the next inter-arrival time.

We shall see that the above defined property of minimal repair is satisfied by the non-homogenous Poisson process (NHPP) to be defined below. This process has been extensively applied in reliability engineering, in particular in studying software reliability modeling (see Chap. 4 of the book by Xie [33] or, more recently, the paper by Pham and Zhang [24] and references therein).

Definition 3.6 (*Non-Homogenous Poisson Process (NHPP)*) Let $N(t)$ as above be the number of failures of the system through time t. The basic probabilistic assumptions in the NHPP are:

(i) $N(0) = 0$;
(ii) $\{N(t), t \geq 0\}$ has the *independent increments* property, which means that for any $a < b \leq c < d$, $N(a, b)$ and $N(c, d)$ are independent random variables;
(iii) $P\{N(t + \varepsilon) - N(t) = 1\} = \mu(t)\,\varepsilon + o(\varepsilon)$;

(iv) $P\{N(t + \varepsilon) - N(t) \geq 2 \} = o(\varepsilon)$, therefore, the probability of simultaneous failures is zero;

where $\mu(t)$ is called the intensity function of the NHPP and corresponds to the unconditional intensity, defined in (3.3)

Let $M(t)$ be the mean function of the NHPP $\{N(t); t \geq 0\}$, as defined in (3.1). Then, it can be deduced that for any given $t \geq 0$ the random variable $N(t)$ follows a Poisson distribution with parameter $M(t)$, that is, the probability that $N(t)$ is a given integer n is expressed as

$$P\{N(t) = n\} = \frac{(M(t))^n}{n!} e^{-M(t)}; \, n = 0, 1, 2, \ldots . \tag{3.8}$$

With $M(t)$ and $\mu(t)$ as defined in the previous section, in general, the probability that the number of failures in an interval $(s, s + t]$, for any $s, t > 0$ is a given n is expressed as

$$P\{N(s, s+t) = n\} = \frac{[M(s+t) - M(s)]^n}{n!} \exp[-(M(s+t) - M(s))], \tag{3.9}$$

for $n = 0, 1, 2, \ldots$

One may characterize $\{N(t); t \geq 0\}$ as an ordered list $S_1 < S_2 < \ldots$ of event times. The stochastic process is specified completely by the joint distribution of $N(t_1) \leq N(t_2) \leq \ldots \leq N(t_n)$ for any sequence $t_1 \leq t_2 \leq \ldots \leq t_n$ and $n \geq 0$. This is equivalent to specifying the joint distribution of the sequence $\{S_1 < S_2 < \ldots \}$. From expressions (3.8) and (3.9) and taking into account the independent increments property, we can get the following

$$P\{S_1 > s\} = P\{N(s) = 0\} = e^{-M(s)} = \exp\left[-\int_0^s m(u)du\right] = \exp\left[-\int_0^s \mu(u)du\right].$$

$$\tag{3.10}$$

So, the first event time of the process is a random variable $T_1 = S_1$ with hazard function $\lambda(t) = \mu(t)$. Let us denote with F its corresponding distribution function. Moreover,

$$P\{S_2 > s_1 + t | S_1 = s_1\} = P\{N(s_1, s_1 + t) = 0\} = e^{-(M(s_1+t)-M(s_1))}$$

$$= \exp\left[-\int_{s_1}^{s_1+t} \mu(u)du\right], \tag{3.11}$$

and, in general we have

$$P\{S_{j+1} > s_j + t | S_j = s_j\} = P\{N(s_j, s_j + t) = 0\} = e^{-(M(s_j+t)-M(s_j))}$$

$$= \exp\left[-\int_{s_j}^{s_j+t} \mu(u)du\right]. \tag{3.12}$$

So, according to the notation in Definition 3.5 and Eq. 3.6, it can be deduced that the distribution of the inter-arrival times of a NHPP that are $T_j = S_j - S_{j-1}$ for $j = 1, 2, \ldots$ and $S_0 = 0$, is expressed as follows

$$P\{T_{j+1} > t | S_j = s_j\} = \bar{F}_{s_j}(t), \qquad (3.13)$$

where F, as above, is the distribution function of the first event time of the NHPP, i.e. $T_1 = S_1$.

The immediate conclusion is that the NHPP satisfies the defining property for minimal repair in (3.7). Hence, the event times of an NHPP can be interpreted as times where the system is minimally repaired. Furthermore, the conditional intensity function $\gamma(t)$ of the counting process of failures equals the unconditional intensity function $\mu(t)$, which is also the rocof $m(t)$, and also gives the hazard function of the random variable that represents the first time to failure of the system T_1, so that we have

$$\gamma(t) = \mu(t) = m(t) = \lambda(t). \qquad (3.14)$$

In the sequel, we will use for this function the notation $\mu(t)$.

One might think that non-homogeneous models are far less tractable than stationary models. However, the difficulty inherent in the renewal function $M(t)$ is noticeable, that is, the expected number of failures in a finite interval $(0, t]$, which should be calculated when perfect repairs are being performed in the system. As we know, in this case, $M(t)$ is given by a sum of infinite convolution functions, $M(t) = \sum_{k=0}^{\infty} F^{(k)}(t)$, where F may be understood as the distribution function of the time to the first failure. Among other considerations, it was explained in Chap. 2 that the renewal function is not available in closed form, except for special cases such as, for example, when F represents an exponential law. The analogous function when the system is being minimally repaired, that is, when the model is a NHPP, is what we called mean function in Definition 3.1, and can be obtained as $M(t) = \int_0^t \lambda(u)\mathrm{d}u = \Lambda(t)$, the cumulative hazard function of the random time to the first failure, see Ascher [5] for further details.

The distribution of a NHPP and, consequently the reliability behavior through time of a minimally repaired system, is characterized completely by the rocof function. The shape of the rocof may exhibit many types of behavior, e.g. decreasing (systems with this property are sometimes called "happy" systems), increasing (for deterioration systems), bathtub shaped, upside-down bathtub shaped to mention a few, but the possibilities are innumerable and could include cyclic behavior with multiple periodicities or marked asymmetric tendencies. In this respect, the work of Krivtsov [19], which considers alternatives to the two models traditionally and more extensively used and which are referred to as the *log-linear* model and *power-law* model, is interesting. He proposes the use of the parametric forms corresponding to a large number of traditional distribution families for the underlying distribution function of the process. However, parametric models cannot generally be flexible enough to cover all the possible shapes of the intensity function and therefore nonparametric procedures turn out to be very appealing. In particular, the main objective of this chapter is to estimate the function μ by using nonparametric techniques from one or more realizations of a NHPP.

3.2 Nonparametric Estimation of the Rate of Occurrence of Failures (Rocof)

The main objective of this chapter is a nonparametric estimation of the rocof. To achieve this, we will consider one or multiple realizations of a NHPP, which means observing a single system or, alternatively, a population of systems of the same characteristics. On the other hand, no assumption is adopted for the functional form of the rocof $\mu(t)$, except that it is a smooth function, smoothness being defined in terms of some properties of differentiability of the function with respect to t.

3.2.1 Smooth Estimation of the Rocof

In this section, failure time observations of repairable systems are considered. The repair times are considered negligible, so they are not included in the analysis of the system reliability. The failures occurring at disjoint intervals are assumed to be independent. There are no simultaneous failures in the system and once the system has failed and been restored to work, its state is exactly the same as it was before failure. The "as bad as old" condition after failure-and-repair or, equivalently, minimal repair maintenance is adopted by the system environment. We observe one system for a test period of $(0,\tau]$ where a total number of n failures have been recorded at times $S_1 < S_2 < \ldots < S_n < \tau$. Thus, in our first approach, the sampling information consists of a single realization of a NHPP and the goal is to obtain an estimate of the rocof function $\mu(t)$.

3.2.1.1 Kernel Estimate

In the case that the data are regularly spaced, that is, failures occur regularly in time, an estimator of the rocof may be built by using kernel functions, that is, defining

$$\widehat{\mu}_h(t) = \frac{1}{h} \sum_{i=1}^{n} w\left(\frac{S_i - t}{h}\right), \tag{3.15}$$

where w is a bounded, non-negative and, in principle, symmetric, density function and h is a bandwidth parameter which quantifies the smoothness of $\widehat{\mu}_h(t)$.

It is worth mentioning some aspects of the estimator. On the one hand, note that its appearance is similar to the typical kernel density estimator except for the normalization factor, which is not necessary in (3.15) since function μ is not expected to integrate to 1, on the contrary, $\int_0^\tau \mu(u)du = E[N(\tau)]$. On the other hand, the sample size, that is, the total number of failures observed in the fixed interval $[0, \tau]$ is random.

Kernel estimation of the intensity function of a counting process was introduced in Ramlau-Hansen [27], where its theoretical properties may be found, as well as in Leadbetter and Wold [21], and Diggle and Marron [14]. In particular, Ramlau-Hansen [27] derives the uniform consistency and asymptotic normality of the estimator of the intensity function of a counting process whose intensity function fits the multiplicative model. In other words, the conditional intensity function is assumed to be of the form $\gamma(t) = \alpha(t)Y(t)$, with α an unknown deterministic function and Y an observable stochastic process (see (3.21) in Sect. 3.2.2). This model was suggested in Aalen [1] and we will return to it further on in the text. The NHPP can be considered a particular case of the multiplicative intensity model. Other analyses of the intensity function of a NHPP or, equivalently, of the rocof of a minimally repaired system have considered the estimator (3.15), see Cowling et al. [13], Phillips [25, 26].

In summary, as in the explanations already given in previous chapters, the kernel-method approximates the value of the rocof at each time point t by using a constant function. The estimations are calculated locally and the bandwidth parameter h controls the size of the local neighborhood, whereas the kernel w assigns weight to each observed failure time. When failures occur regularly through the test period $(0, \tau]$, the kernel estimator gives a satisfactory approximation of the rocof.

3.2.1.2 Local Likelihood Method

It is very common in practical situations that particular periods of time in the life of the system exist where failures arrive with a higher frequency. As a consequence, the sample comprises data that turn out to be very dense in some periods, whereas they become more sparse in other periods of time. In this situation, we explore alternative and potentially better estimation procedures. In this respect, we base our work on that of Wang et al. [32], where the authors adapt the local likelihood method to the particular problem treated here. A logarithmic transformation of the intensity function is considered, that is, it is defined $\varphi(t) = \ln \mu(t)$. The function μ is assumed to be smooth in the sense of certain differentiability properties, in particular it is assumed that the first order derivative exists. Therefore, the function μ can be approximated, using the Taylor expansion, by a linear function of the form

$$\varphi(t) \approx \alpha + \beta(t - t_0), \tag{3.16}$$

where t_0 is a particular time at which we are interested in obtaining an estimation of μ. Obviously, we have $\hat{\mu}(t_0) = \exp(\hat{\alpha})$.

Given a data set consisting of the waiting times of a NHPP (the successive failure times of a system under minimal repair) registered at a test interval like $(0, \tau]$, i.e. $0 = S_0 < S_1 < S_2 < \ldots < S_n < S_{n+1} = \tau$, the likelihood function (see [2]) can be expressed as

$$L = \prod_{i=1}^{n} \mu(S_i) \times \exp\left[-\sum_{i=0}^{n} \int_{S_i}^{S_{i+1}} \mu(u)du\right]$$

$$= \prod_{i=1}^{n} \mu(S_i) \times \exp\left[-\int_{0}^{\tau} \mu(u)du\right]. \tag{3.17}$$

We can express the log-likelihood function in terms of $\varphi(t)$, that is

$$\ell = \ln L = -\int_{0}^{\tau} e^{\varphi(u)}du + \sum_{i=1}^{n} \varphi(S_i). \tag{3.18}$$

A local version of the likelihood can be built by introducing a suitable non-negative weight function w and a bandwidth parameter h. In other words, taking into account the approximation of function φ given in Expression (3.16), the following local weighted log-likelihood function can be obtained

$$\ell_h(t_0) = -\int_{0}^{\tau} \exp[\alpha + \beta(u - t_0)]w_h(u - t_0)du$$

$$+ \sum_{i=1}^{n} [\alpha + \beta(S_i - t_0)]w_h(S_i - t_0), \tag{3.19}$$

where $w_h(\cdot) = w(\cdot/h)/h$ is a kernel function that determines the weights assigned to the observations around t_0 and, in a sense, ensures that the linear approximation of function γ is applied in the neighborhood where it is valid. For a fixed value t_0 in the interval $(0, \tau]$, the local likelihood estimator of $\mu(t_0)$ should be obtained by maximizing (3.19). Suitable values $\hat{\alpha}$ and $\hat{\beta}$ should be provided by numerical methods, since a closed-form solution for the maximization equations does not exist. The numerical solution will lead to the following local estimate

$$\hat{\mu}(t_0) = \exp(\hat{\alpha}). \tag{3.20}$$

The whole curve $\hat{\mu}(t)$, that is, an estimate of the rocof function, can be constructed by running the local log-linear likelihood procedure for t_0 varying in the observation interval $(0, \tau]$.

3.2.2 Bandwidth Selection: Least-Squares Cross-Validation

The kernel estimator of the rocof proposed by Expression (3.15) involves an unknown smoothing parameter that must be estimated. Although the experts on nonparametric statistics admit that in kernel-type estimation problems, the smoothing parameter selection is a far more relevant matter than the choice of the kernel function, there is not much in the literature on the subject that deals with kernel estimation of intensity functions. In order to estimate the smoothing parameter, we consider the least-squares cross-validation method that was introduced by Rudemo [30] and Bowman [10]. The method is a completely automatic (or data driven) procedure for choosing a suitable value for the bandwidth parameter. This method has been extensively applied in other contexts as has kernel density estimation (see [31]). In this section, we refer to the analysis developed by Brooks and Marron [11], who adapt arguments used in kernel density estimation to derive the almost certainly asymptotic optimality of the least-squares cross-validation bandwidth in the intensity function estimation setting. Their reasoning is constructed on the basis of the work by Ramlau-Hansen [27]. These authors consider the NHPP as a particular case of a wider class of counting processes defined by Aalen [1], called *multiplicative intensity models* where the conditional intensity function (see Definition 3.4) is expressed as

$$\gamma(t) = \alpha(t)Y(t),$$
(3.21)

where $\alpha(t)$ is, in general, an unknown non-negative and deterministic function and $Y(t)$ is a *predictable* stochastic process, observable just before t. For their purposes, Brooks and Marron view the rocof of the NHPP within a specific multiplicative structure that they call the simple multiplicative model. In fact, given that a NHPP with intensity μ is observed at a specified interval of time $[0, \tau]$, they consider that

$$\mu(t) = M\alpha(t), \quad t \in [0, \tau],$$
(3.22)

where M is a positive constant, and $\alpha(t)$ is an unknown non-negative deterministic function that integrates to 1 over the observation interval. As a consequence, it is verified that

$$M = \int_0^\tau \mu(t)\mathrm{d}t = E[N(\tau)]$$
(3.23)

and, given that n failures have occurred in the system until τ, the failure times $S_1 < S_2 < \ldots < S_n$, have the same distribution as the order statistics corresponding to n independent random variables with *pdf* $\alpha(t)$ at the interval $[0, \tau]$. The arguments for asymptotic convergence of the Cross-Validation estimate in Brooks and Marron [11] are based on this property. They use arguments previously employed by other authors in the setting of kernel density estimation. It is also worth mentioning that they consider their asymptotic analysis by letting $M \to \infty$, since

$\tau \to \infty$ would result in all the new failures occurring at the right endpoint instead of adding observations everywhere in the interval $[0, \tau]$. Among other things, this approach does not change the shape of the intensity function, which is desirable. So, in their asymptotic study, it is considered that $\alpha(t) = \mu(t)/M$ is a fixed function when $M \to \infty$. Thus, a sequence of intensity functions such as $\mu_r(t) = M_r\, \alpha(t)$ is considered and for each, the corresponding kernel estimate is constructed, $\widehat{\mu}_{r,h}(t)$. According to the results in Ramlau-Hansen [27] this sequence of estimators is asymptotically normal and uniformly consistent as $M_r \to \infty$, $h \to 0$ and $hM_r \to \infty$. Moreover, Brooks and Marron show that the cross-validation method is almost certainly asymptotically optimal for intensity estimation.

More specifically, these authors adopt as a measure of the estimation error the integrated squared error (ISE), which is defined in the context of intensity functions as

$$\text{ISE}(h) = \int_0^\tau (\widehat{\mu}_h(t) - \mu(t))^2 dt, \tag{3.24}$$

with $\widehat{\mu}_h$ as defined in (3.15). Thus, under mild conditions in the kernel, they show, with probability one,

$$\frac{\text{ISE}(h_{\text{cv}})}{\text{ISE}(h_{\text{opt}})} \to 1, \text{ as } n \to \infty, \tag{3.25}$$

where h_{cv} is the value of the bandwidth obtained by the cross-validation technique (see next paragraph), and h_{opt} is the optimal value of the bandwidth, that is, the one that minimizes the ISE.

We now explain briefly the fundamentals behind the least-squares cross-validation technique adapted to intensity kernel estimation. The ISE can be written

$$\text{ISE}(h) = \int_0^\tau \widehat{\mu}_h(t)^2 dt - 2 \int_0^\tau \widehat{\mu}_h(t)\mu(t) dt + \int_0^\tau \mu(t)^2 dt. \tag{3.26}$$

The last term of (3.26) does not depend on h, and so an ideal choice of the bandwidth parameter (in the sense of minimizing the ISE) would correspond to the choice which minimizes the first two terms

$$\int_0^\tau \widehat{\mu}_h(t)^2 dt - 2 \int_0^\tau \widehat{\mu}_h(t)\mu(t) dt. \tag{3.27}$$

The basic principle of the least-squares cross-validation method is to obtain from the data an estimate of the second term of (3.27), replace the corresponding term of (3.27) and then minimize the resulting expression over h. With this in mind, we define $\widehat{\mu}_{-i}$ as the intensity estimate constructed from all the data points except S_i, i.e. the leave-one-out estimator given by

$$\widehat{\mu}_{-i}(t) = \frac{1}{h}\sum_{j \neq i} w\left(\frac{t - S_j}{h}\right).$$
(3.28)

Now we define the cross-validation score function as suggested in Brooks and Marron [11], that is

$$\mathrm{CV}(h) = \int_0^\tau \widehat{\mu}_h(t)^2 \mathrm{d}t - 2\sum_{i=1}^n \widehat{\mu}_{-i}(S_i),$$
(3.29)

where the score function depends only on the data. Moreover, it can be checked that

$$E\left[\sum_{i=1}^n \widehat{\mu}_{-i}(S_i)\right] = E\left[\int_0^\tau \widehat{\mu}(t)\mu(t)\mathrm{d}t\right],$$
(3.30)

and then, substituting the expression (3.30) in the definition of CV(h) shows that $\mathrm{CV}(h) + \int_0^\tau \mu^2(t)\mathrm{d}t$ is a good approximation of the mean integrated squared error (MISE)

$$\mathrm{MISE}(h) = E\left[\int_0^\tau (\widehat{\mu}_h(t) - \mu(t))^2 \mathrm{d}t\right].$$
(3.31)

Finally, assuming that the minimizer of $E[\mathrm{CV}(h)]$ is close to the minimizer of CV(h), it can be expected Eq. 3.29 will provide a good choice for the bandwidth parameter.

Next, we follow an analogous procedure to the one explained in Silverman [31] in the context of density estimation, for expressing the CV(h) in a more suitable form from a computational point of view. Let us assume that the kernel function w is symmetrical. Making $u = t/h$ in the integral term of (3.29), we have the following

$$\int_0^\tau \widehat{\mu}_h(t)^2 \mathrm{d}t = \int_0^\tau \sum_{i=1}^n h^{-1} w\left(\frac{t - S_i}{h}\right) \times \sum_{j=1}^n h^{-1} w\left(\frac{t - S_j}{h}\right) \mathrm{d}t$$

$$= h^{-1}\sum_{i=1}^n \sum_{j=1}^n \int_0^\tau w(h^{-1}S_i - u)w(u - h^{-1}S_j)\mathrm{d}u$$
(3.32)

$$= h^{-1}\sum_{i=1}^n \sum_{j=1}^n w^{(2)}\left(\frac{S_j - S_i}{h}\right),$$

where $w^{(2)}$ is the convolution of the kernel with itself. We can also express

$$\sum_{i=1}^{n} \widehat{\mu}_{-i}(S_i) = \sum_{i=1}^{n} \left\{ \sum_{j=1}^{n} h^{-1} w\left(\frac{S_j - S_i}{h}\right) - h^{-1} w(0) \right\}$$

$$= \sum_{i=1}^{n} \sum_{j=1}^{n} h^{-1} w\left(\frac{S_j - S_i}{h}\right) - nh^{-1} w(0),$$

(3.33)

and, finally,

$$\mathrm{CV}(h) = h^{-1} \sum_{i=1}^{n} \sum_{j=1}^{n} \{w^{(2)} - 2w\} \left(\frac{S_j - S_i}{h}\right) + 2nh^{-1} w(0)$$ (3.34)

Example System Code 1 from DACS Software Reliability Dataset

In this example, we consider the 136 times between failures of System Code 1 that can be obtained from the DACS (The Data and Analysis Center for Software) homepage. This is part of the reliability data set concerning the software failure data on 16 projects and which was compiled by John Musa of Bell Telephone Laboratories. The particular data set that we deal with (System Code 1) consists of the inter-arrival failure times (given in wall-clock seconds), that is $T_1, T_2, \ldots, T_n$ in our notation, where each T_i represents the elapsed time from the previous failure to the current failure. So we first need to obtain the sequence of successive failure times, that is $S_1 < S_2 < \ldots < S_n$.

We assume that the failures occur according to a NHPP whose rocof function is to be estimated nonparametrically. To do this, an estimator of type (3.15) has been constructed and an appropriate value of the bandwidth parameter has been obtained by the cross-validation procedure explained in Sect. 3.2.2. The results are displayed in Fig. 3.1. The graph also allows to compare this nonparametric fitting to the one

Fig. 3.1 Cross-validation bandwidth selection for kernel estimation of the rocof

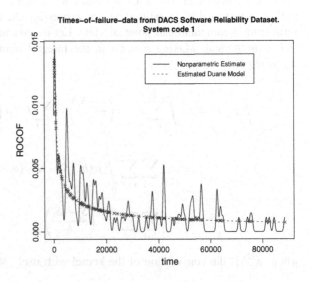

obtained by modeling the rocof according to a power-law or Duane model, which can be expressed as $\mu(t) = (\beta/\eta)(t/\eta)^{\beta-1}$, for suitable $\beta > 0$ and $\eta > 0$.

Remark The nearest neighbor method. The drawback of the least-squares cross-validation bandwidth just proposed is that it is constant all over the observation window and therefore does not account for patterns of failure through the life of the system. To avoid this problem, it might be appropriate to apply alternative procedures such as the k-nearest neighbor method. In such a case, the smoothing parameter is defined in terms of the number of failures, say k, that are considered for the estimation of the rocof at a particular time, t. More specifically, the empirical cumulative distribution function (F_n) associated with the sequence $S_1 < S_2 < \ldots < S_n$ is constructed and the size of the estimation window around a particular time t is determined according to

$$h_k(t) = \inf\left\{h > 0 : F_n\left(t + \frac{h}{2}\right) - F_n\left(t - \frac{h}{2}\right) \geq \frac{k}{n}\right\}. \tag{3.35}$$

The next task is to find an appropriate value for k, which should be done by minimizing a suitable measure of the error such as the ISE. This question is beyond our consideration. A simple procedure related to the above is included in the work of Wang et al. [32], which proposes a hybrid method involving variable bandwidths and the nearest neighbor procedure.

3.2.3 Bootstrap Confidence Bounds for Rocof

When dealing with the problem of estimation of an unknown quantity, one important issue is how to quantify the uncertainty in the estimate. When the objective is to estimate a single numerical value, a confidence interval is constructed to give an idea of the accuracy of a point estimate. In the context of curve-estimation, confidence bands are required to perform a similar role.

In this section, we consider the techniques devised by Cowling et al. [13] and further used in Phillips [25, 26] to achieve this. In the seminal paper by Cowling et al. [13], several bootstrap methods are presented for constructing confidence regions for the intensity function of a NHPP. Further, these methods are applied by Phillips [25, 26] to the estimation of the expected rocof of a repairable system.

Let us first define what is understood by confidence bands.

Definition 3.7 (*Confidence Region*) A connected non-empty random set $\mathcal{B}$ of the rectangle $[0, \tau] \times [0, \infty)$ such that $\mathcal{B} \cap [0, \tau] \times (0, \infty)$ is non-empty for each $t \in [0, \tau]$ is called a *confidence region* for μ over the set $\mathcal{I} \subseteq [0, \tau]$ with coverage $1 - \alpha$ if

$$\Pr\{(t, \mu(t)) \in \mathcal{B} \quad \text{for all} \quad t \in \mathcal{I}\} = 1 - \alpha$$

In the case that $\mathcal{I}$ is a point or a finite set of points, confidence intervals are obtained using the definition above. When $\mathcal{I}$ is considered to be an interval, we derive what is known as a simultaneous confidence band, which may be understood as a collection of confidence intervals for all values of $t \in \mathcal{I}$, constructed to have simultaneous coverage probability of, say $1 - \alpha$. In other words, all the intervals in the collection cover their true value simultaneously with a probability of $1 - \alpha$.

We consider $\mathcal{I}$ to be an interval, so we will construct a simultaneous confidence band which, for the sake of simplicity, will be represented as $\widehat{\mu}(t) \pm B(t)$, and, according to Definition 3.7, must satisfy the condition

$$\Pr\{\widehat{\mu}(t) - B(t) \le \mu(t) \le \widehat{\mu}(t) + B(t), \quad \forall t \in \mathcal{I}\} = 1 - \alpha$$

We then consider percentile-t bootstrap confidence bands as presented in Cowling et al. [13]. Next, we organize the steps to derive confidence bands for the rocof by following the guidelines suggested by these authors. First of all, we reproduce here one of the resampling methods introduced in the paper cited, where bootstrap methods are applied in the context of NHPP. Consequently, we consider a percentile-t procedure to construct confidence regions.

Let $0 < S_1 < S_2 < \ldots < S_n < \tau$ be the failure times observed in the interval $[0, \tau]$.

1. **Resampling Method**

   Conditional on the data, generate n^* from a Poisson distribution with parameter $M(\tau) = \int_0^\tau \mu(u)du$, and draw $S_1^*, S_2^*, \ldots, S_{n^*}^*$ by sampling randomly with replacement n^* times from the data set $S_1, S_2, \ldots, S_n$. This method produces resamples of varying sizes and that may include ties. The procedure is motivated by the fact that, conditional on the event $N(\tau) = n$, the $S_1 < S_2 < \ldots < S_n$ are the ordered statistics of a sample of size n drawn from a distribution with pdf $f(t) = \mu(t)/M(\tau)$. The bootstrap version of the estimator $\widehat{\mu}_h$ defined in Expression (3.15) is then

   $$\widehat{\mu}_h^*(t) = \frac{1}{h} \sum_{i=1}^{n^*} w\left(\frac{S_i^* - t}{h}\right) \tag{3.36}$$

   One reason for choosing this resampling procedure over others is that, in this case, it is true that, $E_*\left[\widehat{\mu}_h^*\right] = \widehat{\mu}_h$, which simplifies some expressions, as will be verified below. Here, the subscript $*$ indicates that the expectation is calculated conditional to the data set.

2. **Bootstrap Approximation**

   The method considered here for constructing simultaneous confidence bands for μ is based on the bootstrap approximation

   $$\begin{aligned} X^*(t) &= \frac{\widehat{\mu}_h^*(t) - E_*\left[\widehat{\mu}_h^*(t)\right]}{\sqrt{\mathrm{Var}(\widehat{\mu}_h(t))}} \\ &= \frac{\widehat{\mu}_h^*(t) - \widehat{\mu}_h(t)}{\sqrt{\mathrm{Var}(\widehat{\mu}_h(t))}}, \quad t \in [0, \tau]. \end{aligned} \tag{3.37}$$

of

$$X(t) = \frac{\widehat{\mu}_h(t) - E[\widehat{\mu}_h(t)]}{\sqrt{\text{Var}(\widehat{\mu}_h(t))}}, \quad t \in [0, \tau]. \tag{3.38}$$

The validity of this approximation is established in Cowling et al. [13] where properties of the estimator $\widehat{\mu}_h$ such as variance and bias are also obtained. Specifically, the authors show that the variance is approximately proportional to μ. Hence, instead of (3.37), the following expression is used as a bootstrap approximation of (3.38)

$$X^*(t) = \frac{\widehat{\mu}_h^*(t) - \widehat{\mu}_h(t)}{\sqrt{\widehat{\mu}_h^*(t)}}, \quad t \in [0, \tau]. \tag{3.39}$$

3. **Percentile-t Confidence Bands**

Let us consider here a procedure to construct confidence bands for μ of a width proportional to the standard deviation. Specifically, let us define C_α such that

$$\text{Pr}_*\{|X^*(t)| \leq C_\alpha \ \forall t \in \mathcal{I}\} = 1 - \alpha \tag{3.40}$$

where, as above, the subscript * indicates that the probability is calculated conditional to the data set. Thus, for any $t \in [0, \tau]$, consider the limits given by

$$\mathcal{B}_1 = \max\{0, \widehat{\mu}_h(t) - C_\alpha\sqrt{\widehat{\mu}_h(t)}\} \quad \text{and} \quad \mathcal{B}_2 = \widehat{\mu}_h(t) + C_\alpha\sqrt{\widehat{\mu}_h(t)}, \tag{3.41}$$

and define the region

$$\mathcal{B} = [0, \tau] \times (\mathcal{B}_1, \mathcal{B}_2). \tag{3.42}$$

As can be appreciated, at each $t \in [0, \tau]$, the width of the corresponding interval is proportional to $\sqrt{\widehat{\mu}_h(t)}$, thus reflecting the variability of the point estimate.

Example U.S.S. *Grampus* Number 4 Main Propulsion Diesel Engine

The data set considered in this example can be found on page 395 of Meeker and Escobar [22]. The data consist of the times (in thousands of operating hours) of unscheduled maintenance actions, caused by system failure or imminent failure, for the number 4 diesel engine of the U.S.S. *Grampus*, up to 16,000 h of operation. We have constructed a confidence region for this data set in Fig. 3.2, where the estimated rocof and the upper and lower limits of the confidence region are given for the time interval where there were observed failures.

3.3 Recurrence Data: The Multiple Realization Setup

In this section we deal with the estimation of the intensity function of a NHPP in the case that several realizations of the process are available. In the reliability setting language, we assume that two or more identical systems are being

Fig. 3.2 Bootstrap
percentile-t confidence bands

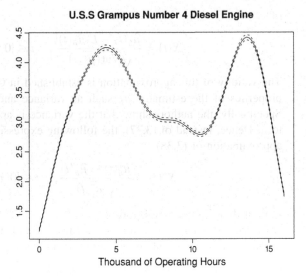

U.S.S Grampus Number 4 Diesel Engine

Thousand of Operating Hours

independently observed through time. We assume that we have failure data from a collection of k identical systems. The occurrence of failures in each can be modeled as a non-homogeneous Poisson process with unknown intensity function $\mu(t)$. So, the observation of each individual system provides a realization of a NHPP with intensity $\mu(t)$. Let $N_j(t)$ denote the cumulative number of failures for system j before time t. We follow each system in the sample for a period of time that is assumed to be independent of the system's history. During the time interval $[0, \tau_j]$ the jth system exhibits failures at times $S_{j1} < S_{j2} < \ldots < S_{jn\_j} \leq \tau_j$, for each $j = 1, 2, \ldots, k$. To illustrate we present the following example.

Example Hydraulic system of load-haul-dump (LHD) machines (Kumar and Klefsjö [20])

Figure 3.3 shows the observed failure process of the hydraulic subsystems of a sample of load-haul-dump machines, which have been used to pick up ore or waste rock from mining points and dump it into trucks or ore passes. The data represented in Fig. 3.3 appear in the Appendix of Kumar and Klefsjö [20] and consist of the times between successive failures (in hours, and excluding repair or down times) of the hydraulic systems of the LHD machines that were selected for study by these authors.

In their analysis, the authors develop several methods in order to detect the presence of trends in the times between failures of the hydraulic systems. In particular, they show that these times are neither independent nor identically distributed and, in fact, they check graphically that the hydraulic system times between failures are decreasing and in addition that a NHPP with power-law intensity function provides a good fit. The graph in Fig. 3.3 was created by using the package *gcmrec* [16] of R, which was designed to handle recurrent event data.

Fig. 3.3 Hydraulic systems
event plot

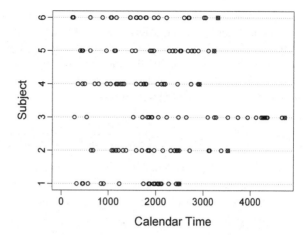

As in (3.17), we can construct in a similar way a partial likelihood function based on each realization of the process L_j, that is

$$L_j(\mu) = \prod_{i=1}^{n_j} \mu(S_{ji}) \times \exp\left[-\int_0^{\tau_j} \mu(u)du\right], \tag{3.43}$$

for $j = 1, 2, \ldots, k$. So the joint likelihood given the event time observations is

$$
\begin{aligned}
L(\mu) &= \prod_{j=1}^{k} L_j(\mu) \\
&= \left[\prod_{j=1}^{k}\prod_{i=1}^{n_j} \mu(S_{ji})\right] \times \exp\left[-\sum_{j=1}^{k}\int_0^{\tau_j} \mu(u)du\right].
\end{aligned} \tag{3.44}
$$

The naive nonparametric maximum-likelihood estimate for μ, which gives an infinite value to the likelihood, is

$$\widehat{\mu}(t) = \sum_{j=1}^{k}\sum_{i=1}^{n_j} \delta(t - S_{ji}), \tag{3.45}$$

where δ is the Dirac delta function. In the existing literature, other more useful nonparametric estimates have been derived. We will now retrieve some of these.

3.3.1   Estimation Procedures with Observed Failure Times

Bartoszyński et al. [8] develop a study of the kinetic mechanism for the spread of cancer. They model the appearance of metastases as a non-homogeneous Poisson process with intensity μ whose functional form is not specified beforehand, rather the authors let the data determine its shape. So they consider nonparametric estimation procedures for approximating the value of μ, in particular they present a window-type estimate for the intensity function. We adapt their reasoning to estimate the rocof of a recurrence process of failures in the context of reliability systems. Define the following estimator

$$\widehat{\mu}_h(t) = \frac{1}{h} \frac{\# \text{failures in}(t - h/2, t + h/2)}{\# \text{systems under observation at time } t} \tag{3.46}$$

Let $\delta_j(t) = I[\tau_j \geq t]$ denote the *at-risk* indicator corresponding to the jth system of the sample. In other words, $\delta_j(t) = 1$ if system j is still being observed at time t and $\delta_j(t) = 0$ otherwise. Let $\delta.(t) = \sum_{j=1}^{k} \delta_j(t)$ be the size of the risk set at time t. The expression in (3.46) can be written as

$$\widehat{\mu}_h(t) = \frac{1}{h} \sum_{j=1}^{n} \frac{\delta_j(t)}{\delta.(t)} \sum_{i=1}^{n_j} I\big[S_{ji} \in (t - h/2, t + h/2)\big]. \tag{3.47}$$

This is a kernel-type estimate with kernel function

$$w(\cdot) = I[|u| \leq 0.5]. \tag{3.48}$$

Then, defining $w_h(\cdot) = w(\cdot/h)/h$, (3.47) can be written

$$\widehat{\mu}_h(t) = \sum_{j=1}^{n} \frac{\delta_j(t)}{\delta.(t)} \sum_{i=1}^{n_j} w_h\big(S_{ji-t}\big), \tag{3.49}$$

and other kernel functions w, different from (3.48), could be chosen. However, since in kernel estimation setting it is generally accepted that the choice of the kernel shape has a relatively small importance (when compared to the choice of the bandwidth parameter h), we will limit ourselves to the function given in (3.48). Note that the expression in (3.49) is obtained as the average of the kernel estimates obtained by considering each realization of the NHPP separately, see (3.15).

As is established in Bartoszyński et al. [8], if the number of systems censored in the interval $(t - h/2, t + h/2)$ is negligible relative to $\delta.(t)$ and if μ is twice differentiable at t, then, given the size of the risk set, the following expression can be derived for the mean squared error (MSE)

$$\text{MSE}_{\delta.}(t, h) = E_{\delta.}\left[(\widehat{\mu}_h(t) - \mu(t))^2\right]$$
$$= \frac{\mu(t)}{\delta.} + \frac{h^2 \mu^4(t)}{4} + O\big(\delta.(t)^{-1}\big) + O(h^3). \tag{3.50}$$

This expression can be minimized, leading to the following value of the bandwidth, which will be considered optimal in terms of the MSE

$$h_{\text{opt}}(t) = \left(\frac{2}{\delta.(t)}\right)^{1/3} \frac{1}{\mu(t)}. \tag{3.51}$$

So, we achieve a value of the bandwidth parameter that depends on the intensity function that is the quantity we want to estimate. We adapt the iterative algorithm presented in Bartoszyński et al. [8] to estimate $\mu(t)$ at each failure time. The algorithm may be summed up in a few steps.

Algorithm Rocof Estimation with Recurrent Failure Times

Step 1. Consider as initial value of the rocof a constant estimate given by the total number of failures divided by the total duration of the observation period, that is

$$\widehat{\mu}_0(t) = \frac{\sum_{j=1}^{k} n_j}{\max\{\tau_j : j = 1, 2, \ldots k\}} \tag{3.52}$$

Step 2. Given a previous estimate $\widehat{\mu}_l$, calculate h_{l+1} by (3.51) as

$$h_{l+1}(t) = \left(\frac{2}{\delta.(t)}\right)^{1/3} \frac{1}{\widehat{\mu}_l(t)}. \tag{3.53}$$

Step 3. Update the estimation of the rocof by means of $\widehat{\mu}_{l+1}$ calculated as in (3.49) with h_{l+1}, and stop if this updated value is close enough (according to a suitable convergence criterion) to the one obtained at the previous iteration, otherwise go to Step 2.

3.3.2 Estimation Procedures with Aggregated Data

In many practical situations of systems maintenance, individual records of failure times are not available. Instead, aggregate information over fixed intervals of time is recorded. This section considers the problem of estimating the rocof function when data are in this form. To be more precise, let us assume that k realizations of a NHPP are being considered in an interval of time $[0, \tau]$. Moreover, let us assume that information about the number of failures is only available at fixed increments of time of magnitude $\Delta > 0$. Therefore, for each system in the sample, the number of failures in each interval of the form $[(i-1)\Delta, i\Delta)$ is reported, for $i = 1, 2, \ldots, m = \tau \setminus \Delta$, where the symbol $\cdot \setminus \cdot$ represents integer division. In other words, the data consist of the number of events of each realization of the process in each interval of fixed size, say Δ.

We reproduce here the estimation procedure proposed by Henderson [17]. Let $N_j(t_1, t_2)$ denote the number of failures suffered by the jth system in a generic interval $[t_1, t_2)$. Let us consider a partition in the global interval $[0, \tau]$ in subintervals of width Δ and, for any $0 \le t < \tau$ define $i(t) = (t \setminus \Delta) * \Delta$ so that t belongs to the subinterval $[i(t), i(t) + \Delta)$. Henderson [17] gets round the problem by first defining an estimate for the cumulative rate function $M(t) = \int_0^t \mu(u)du$. He starts by considering an approximation of this function by linearly interpolating into each interval of type $[(i-1)\Delta, i\Delta)$, for $i = 1, 2, ..., m$. This leads to the following function

$$M_0(t) = M(i(t)) + \frac{M(i(t) + \Delta) - M(i(t))}{\Delta}(t - i(t)) \qquad (3.54)$$

with $i(t)$ defined as above. The author describes an estimator of this function by means of

$$\widehat{M}_0(t) = \frac{1}{k}\sum_{j=1}^{k} N_j(0, i(t)) + \frac{\sum_{j=1}^{k} N_j(i(t), i(t)\Delta)}{k\Delta}(t - i(t)). \qquad (3.55)$$

Some properties of this estimator are also obtained. On the one hand, it is proven that the estimator converges to the true value of $M(t)$ at the limits of the partition of $[0, \tau)$ in Δ-intervals. The rate of convergence is established as $k^{-1/2}$, which obviously depends on the number of realizations. Finally, the author considers an estimator of the cumulative rate similar to (3.55) but now $\Delta = \Delta_k$ is allowed to depend on k. An asymptotic criterion in terms of the number of realizations is used and, under mild conditions over the intensity function, the uniformly strong consistency and asymptotic normality of the estimator when $\Delta_k \to 0$ with $k \to \infty$ is shown. From this starting point, at each $0 \le t < \tau$ the 'aggregated' rate function is defined as the slope of the local interpolation of the cumulative rate, that is

$$\mu_0(t) = \frac{M(i(t) + \Delta) - M(i(t))}{\Delta}, \qquad (3.56)$$

which is obviously constant in each Δ-interval. In parallel, the corresponding estimator of the rocof is then defined as the slope of the linear estimate of the cumulative rate function, so

$$\widetilde{\mu}_0(t) = \frac{\sum_{j=1}^{k} N_j(i(t), i(t)\Delta)}{k\Delta}. \qquad (3.57)$$

This estimator exhibits similar large-sample properties as those for the cumulative rate function estimator.

Example Hydraulic System LHD Machines (continued)

Now we give an illustration of the procedure explained in Sect. 3.3.1 for constructing a nonparametric estimate of the rocof when the observed failure times corresponding to several realizations of an NHPP are available. To do this, we base ourselves on the data set displayed in Fig. 3.3, which presents the observed calendar times (in hours), excluding repair or down times, of the successive failures of the hydraulic subsystems of a sample of load-haul-dump machines used in underground mines in Sweden. See Kumar and Klefsjö [20] for further details. In their study, these authors fit a power-law model to this data set and conclude that

Fig. 3.4 Hydraulic systems rocof estimate

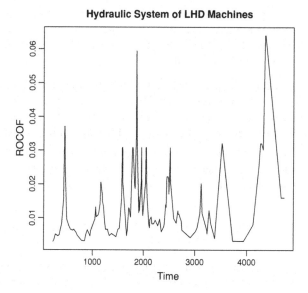

the system is deteriorating with time. This assertion is also admissible by considering the nonparametric approach as can be appreciated from Fig. 3.4. The advantage of the nonparametric estimated is that local features can be highlighted, like the increase near 2,000 h, which is not evident from the power-law estimate.

3.4 Nonparametric Maximum-Likelihood Estimation of the Rocof Under Monotonicity Constraints

3.4.1 One Realization Setup

Let $S_1 < S_2 < \ldots < S_n$, as always, be the recorded failure times of a system, which is being observed in an interval of time $(0, \tau]$. The failures occur randomly in time according to a NHPP with unknown rocof $\mu(t)$. As explained in previous sections, the likelihood function associated with the observed data is

$$L(\mu) = e^{-M(S_n)} \times \left[\prod_{i=1}^{n} \mu(S_i) \right], \qquad (3.58)$$

where $M(t) = \int_0^t \mu(u)\mathrm{d}u$. We are interested in the optimization problem described next [9].

Problem 1 (*Find a function μ which maximizes (3.58), for fixed $S_1 < S_2 < \ldots < S_n$, subject to the following restrictions*)

$$0 \le \mu(t) \le C, \quad \mu(t) \text{ is non-decreasing.}$$

As pointed out in Boswell [9], the upper bound C is needed. Otherwise *Problem 1* would not have a solution since $\mu(t)$ could be set equal to some positive number for each $0 < t < S_n$ and $\mu(S_n)$ could be defined as whatever size was wanted, making (3.58) arbitrarily large.

It can be appreciated from (3.58) that the maximum value depends on how the rocof is defined for $t \neq S_i$ for $i = 1, 2, \ldots, n$ only through the exponential term, so, with the values $\mu(S_i)$ fixed, the problem is reduced to determining the intermediate values of function μ which maximize this factor. In other words, a function μ must be found to minimize the integral $M(S_n) = \int_0^{S_n} \mu(u) du$. If this integral is understood as the area below the curve μ, the solution is given by a step function such as the following

$$\mu(t) = \begin{cases} 0 & \text{if} \quad 0 \leq t < S_1 \\ \mu(S_i) & \text{if} \quad S_i \leq t < S_{i+1}, \quad i = 1, 2, \ldots, n-1 \\ \mu(S_n) = C & \text{if} \quad S_n \leq t. \end{cases} \quad (3.59)$$

Let us assume that a constant C is selected beforehand. Thus, to solve *Problem 1* we must find values $0 \leq \widehat{\mu}_1 \leq \widehat{\mu}_2 \leq \ldots \leq \widehat{\mu}_{n-1}$ which maximize

$$\exp\left[-\sum_{i=1}^{n-1} s_i \mu_i\right] \times \left[\prod_{i=1}^{n-1} \mu_i\right], \quad (3.60)$$

and take $\widehat{\mu}(S_i) = \widehat{\mu}_i$, with $s_i = S_{i+1} - S_i$, for $i = 1, 2, \ldots, n-1$. Theorem 2.2 in Boswell [9] gives the solution to the problem and proves that the maximum-likelihood estimate of $\mu(t)$ over $(0, S_n)$ for $\mu(t)$ satisfying the conditions specified in *Problem 1* is

$$\widehat{\mu}(t) = \begin{cases} 0 & \text{if} \quad 0 \leq t < S_1 \\ \min\{C, \widehat{\mu}_i\} & \text{if} \quad S_i \leq t < S_{i+1}, \quad i = 1, 2, \ldots, n-1, \\ C & \text{if} \quad S_n \leq t \end{cases} \quad (3.61)$$

where

$$\widehat{\mu}_i = \max_{1 \leq a \leq i} \min_{i < b \leq n} \left\{ \frac{(b-a)}{(S_b - S_a)} \right\}.$$

3.4.2 Multiple Realization Setup

In this section we consider the case of several identical systems being observed under the same conditions. With time, failures are recorded in each system according to a NHPP with rocof $\mu(t)$. A situation similar to the one described in Fig. 3.3 is being considered. So, a new derivation of the maximum-likelihood of a monotonic rocof is outlined. This point estimator is presented in Zielinski et al. [34] and is a direct generalization of the one obtained in the single-system realization setup presented in the previous section.

The data consist of k independent realizations of a NHPP with intensity $\mu(t)$. For the jth system in the sample, failures are registered at times $0 < S_{j1} < S_{j2} < \ldots < S_{jn_j} \leq \tau_j$. To position us in the more general situation, let us assume that the censoring times $\tau_1, \tau_2, \ldots, \tau_k$ are independent random variables with joint density given by $g(\tau_1, \tau_2, \ldots, \tau_k)$. So, the likelihood for the observed data is the product of the individual likelihoods, see (3.58) that is

$$L(\mu) = \left[\prod_{j=1}^{k} \prod_{i=1}^{n_j} \mu(S_{ji}) \right] \times \exp\left[-\sum_{j=1}^{k} \int_0^{\tau_j} \mu(s)ds \right] \times g(\tau_1, \tau_2, \ldots, \tau_k). \quad (3.62)$$

Since the censoring times are independent of the underlying Poisson process and also, therefore, of the rocof function, when maximizing the likelihood function in μ, we can ignore the corresponding term. So, defining $\tau = \max\{\tau_1, \tau_2, \ldots, \tau_k\}$, the maximization problem can now be established as given in the following.

Problem 2 (*Find a function μ which maximizes*)

$$\ell^*(\mu) = \left[\sum_{j=1}^{k} \sum_{i=1}^{n_j} \ln \mu(S_{ji}) \right] - \left[\sum_{j=1}^{k} \int_0^{\tau_j} \mu(s)ds \right] \quad (3.63)$$

over the interval $[0, \tau]$, *subject to the following restrictions*

$$0 \leq \mu(t) \leq C, \quad \mu(t) \text{ is non-decreasing.}$$

Reasoning as we did in the one realization case, the solution is a step function closed on the left and with jumps, at the most, at the observed times of failure. Let us arrange the event times into a single list,

$$0 < S_1 < S_2 < \cdots < S_N = \tau,$$

where N is the total number of different event times between the k realizations of the process. Let us define the following quantities suggested in Bartoszyński et al. [8] and Zielinski et al. [34],

$$N_r = \text{\# of failures recorded at time } S_r;$$

$$\Delta_r = \sum_{j=1}^{k} \max\left\{ 0, \min\left\{ (\tau_j - S_r), (S_{r+1} - S_r) \right\} \right\};$$

$$\mu_r = \text{value of } \mu(\cdot) \text{ at } S_r;$$

for $r = 1, 2, \ldots, N$. According to the above comments, *Problem 2* can be reduced to find values $\widehat{\mu}_1 \leq \widehat{\mu}_2 \leq \cdots \leq \widehat{\mu}_N = C$ which maximize

$$\sum_{r=1}^{N} (N_r \ln \mu_r - \Delta_r \mu_r) \quad (3.64)$$

over the interval $[0, \tau]$. Bartosziński et al. [8] show that the solution to *Problem 2* exists and is unique. They provide a constructive proof of its existence but they do not give the solution explicitly. On the other hand, Zielinski et al. [34] obtain in closed form the solution of *Problem 2* which we represent as

$$\hat{\mu}_r = \min_{1 \le a \le r} \max_{r \le b \le N} \frac{\sum_{i=a}^{b} N_i}{\sum_{i=a}^{b} \Delta_i}, \tag{3.65}$$

for $r = 1, 2, \ldots, N$. Thus, it can be written

$$\hat{\mu}(t) = \begin{cases} 0 & \text{if } 0 \le t < S_1 \\ \hat{\mu}_r & \text{if } S_r \le t < S_{r+1}, \ r = 1, 2, \ldots, N-1 \\ C & \text{if } S_N \le t. \end{cases} \tag{3.66}$$

Example U.S.S. *Halfbeak* Number 4 Main Propulsion Diesel Engine

To illustrate the procedure for nonparametric estimation of the rocof under monotonicity constraints, we have considered again a data set included in Meeker and Escobar [22]. In particular, we make use of the data included in Table 16.4, page 415 of [22] where the unscheduled maintenance actions for the U.S.S. *Halfbeak* number 4 main propulsion diesel engine over 25,518 h of operation are specified. In this case, according to the analysis carried out by the authors, we assume that the system is deteriorating with time, so we consider as valid the hypothesis that the rocof is a non-decreasing function. With this monotonicity constraint, we have constructed the nonparametric estimate as explained in Sect. 3.4.1, and we have obtained the results that are showed in Fig. 3.5, where the

Fig. 3.5 Rocof estimate under monotonicity constraints

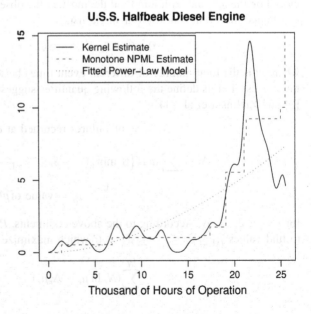

U.S.S. Halfbeak Diesel Engine

— Kernel Estimate
--- Monotone NPML Estimate
······ Fitted Power–Law Model

Thousand of Hours of Operation

power-law model fitted in Meeker and Escobar [22] and the (unconstrained) kernel estimate are also presented.

References

1. Aalen O (1978) Nonparametric inference for a family of counting processes. Ann Stat 6(4):701–726
2. Andersen P, Borgan O, Gill R, Keiding N (1993) Statistical models based on counting processes. Springer, New York
3. Ascher H (1968) Evaluation of a repairable system reliability using 'bad-as-old' concept. IEEE Trans Reliab 17:103–110
4. Ascher H (2007a) Repairable systems reliability. In: Ruggeri F, Kenett R, Faltin FW (eds) Encyclopedia of statistics in quality and reliability. Wiley, New York
5. Ascher H (2007b) Different insights for improving part and system reliability obtained from exactly same DFOM 'failure numbers'. Reliab Eng Syst Saf 92:552–559
6. Ascher H, Feingold H (1984) Repairable systems reliability: modeling, inference, misconceptions and their causes. Marcel Dekker, New York
7. Aven T (2007) General minimal repair models. In: Ruggeri F, Kenett R, Faltin FW (eds) Encyclopedia of statistics in quality and reliability. Wiley, New York
8. Bartoszyński R, Brown BW, McBride M, Thompson JR (1981) Some nonparametric techniques for estimating the intensity function of a cancer related nonstationary Poisson process. Ann Stat 9(5):1050–1060
9. Boswell MT (1966) Estimating and testing trend in a stochastic process of Poisson type. Ann Math Stat 37(6):1564–1573
10. Bowman AW (1984) An alternative method of cross-validation for the smoothing of density estimates. Biometrika 71:353–360
11. Brooks MM, Marron JS (1991) Asymptotic optimality of the least-squares cross-validation bandwidth for kernel estimates of density functions. Stoch Process Appl 38:157–165
12. Chiang CT, Wang M-C, Huang CY (2005) Kernel estimation of rate function for recurrent event data. Scand J Stat 32(1):77–91
13. Cowling A, Hall P, Phillips MJ (1996) Bootstrap confidence regions for the intensity of a Poisson point process. J Am Stat Assoc 91:1516–1524
14. Diggle PJ, Marron JS (1988) Equivalence of smoothing parameter selectors in density and intensity estimation. J Am Stat Assoc 83:793–800
15. Finkelstein M (2008) Failure rate modelling for reliability and risk. Springer, London
16. González JR, Slate EH, Peña EA (2009) gcmrec: general class of models for recurrent event data http://www.r-project.org
17. Henderson SG (2003) Estimation for nonhomogeneous Poisson process from aggregated data. Oper Res Lett 31:375–382
18. Hollander M, Samaniego FJ, Sethuraman J (2007) Imperfect repair. In: Ruggeri F, Kenett R, Faltin FW (eds) Encyclopedia of statistics in quality and reliability. Wiley, New York
19. Krivtsov VV (2006) Practical extensions to NHPP application in repairable system reliability analysis. Reliab Eng Syst Saf 92:560–562
20. Kumar U, Klefsjö B (1992) Reliability analysis of hydraulic systems of LHD machines using the power law process model. Reliab Eng Syst Saf 35:217–224
21. Leadbetter MR, Wold D (1983) On estimation of point process intensities, contributions to statistics: essays in honour of Norman L. Johnson, North-Holland, Amsterdam
22. Meeker WQ, Escobar LA (1998) Statistical methods for reliability data. Wiley, New York
23. Nelson W (2003) Recurrent events data analysis for product repairs, disease recurrences, and other applications. (ASA-SIAM series on statistics and applied probability)

24. Pham H, Zhang X (2003) NHPP software reliability and cost models with testing coverage. Eur J Oper Res 145(2):443–454
25. Phillips MJ (2000) Bootstrap confidence regions for the expected ROCOF of a repairable system. IEEE Trans Reliab 49:204–208
26. Phillips MJ (2001) Estimation of the expected ROCOF of a repairable system with bootstrap confidence region. Qual Reliab Eng Int 17:159–162
27. Ramlau-Hansen H (1983) Smoothing counting process intensities by means of kernel functions. Ann Stat 11:453–466
28. Rausand M, Høyland A (2004) System reliability theory models, statistical methods, and applications. Wiley, New York
29. Rigdon SE, Basu AP (2000) Statistical methods for the reliability of repairable systems. Wiley, New York
30. Rudemo M (1982) Empirical choice of histograms and kernel density estimators. Scand J Stat 9:65–78
31. Silverman BW (1986) Density Estimation for Statistics and Data Analysis. Chapman and Hall, London
32. Wang Z, Wang J, Liang X (2007) Non-parametric estimation for NHPP software reliability models. J Appl Stat 34(1):107–119
33. Xie M (1991) Software reliability modelling. World Scientific, Singapore
34. Zielinski JM, Wolfson DB, Nilakantan L, Confavreux C (1993) Isotonic estimation of the intensity of a nonhomogeneous Poisson process: the multiple realization setup. Can J Stat 21(3):257–268

Chapter 4
Models for Imperfect Repair

4.1 Introduction

In Chaps. 2 and 3, we have considered systems that are either replaced or minimally repaired at failures. The standard models for those cases are, respectively,
the renewal process (RP) and the nonhomogeneous Poisson process (NHPP).
The RP model is suitable if the system at failures is perfectly repaired or
completely replaced, while the NHPP is appropriate if only a minor part of the
system is repaired or replaced at failures. For many applications, it is, however,
more reasonable to model the repair action by something in between the two given
extremes of perfect and minimal repair.

We will consider two main approaches for modeling such failure and repair
processes. The first approach is via the modeling of conditional intensities, which
combines in some sense the features of the "extreme" repair processes, RP and
NHPP. The second approach is via the notion of effective ages, where the idea is
that repairs may result in improved system behavior equivalent to a certain
reduction in time since the system was new.

The basic notation and basic concepts for counting processes used in the present
chapter follow the definitions given in Sect. 3.1.2.

4.1.1 Models for Imperfect Repair Based on Conditional Intensities

Cox [9] introduced a class of point processes called the modulated renewal
process. An interesting feature of the modulated renewal process is that it contains
both the NHPP and the RP as special cases. In particular, this makes it possible to
choose between a minimal and perfect repair model by use of statistical (likelihood
ratio) tests for the parameters.

M. L. Gámiz et al., *Applied Nonparametric Statistics in Reliability*,
Springer Series in Reliability Engineering, DOI: 10.1007/978-0-85729-118-9_4,
© Springer-Verlag London Limited 2011

The modulated renewal process can be viewed as an intensity-based model similar to the Cox regression model. A basic difference is, however, that it includes the *time since last failure* as a covariate. Such a covariate is called dynamic. Another example of a dynamic covariate that may enter a conditional intensity function is the number of events before time t, $N(t-)$. This type of covariates also enters the model by Peña and Hollander [29], which will be considered in Sect. 4.3.2 of the present chapter. In addition, their model involves the effective age of the system under consideration, to be described in Sect. 4.1.2 below. A nice discussion of dynamic covariates in models for recurrent events can be found in the book by Aalen et al. [1].

A related class of processes called inhomogeneous gamma processes was suggested by Berman [6]. Berman motivated the inhomogeneous gamma process by first considering the point process of observing every κth event of an NHPP, where κ is a positive integer. Recall that for an NHPP with events at times S_1, S_2, ... and with cumulative intensity $\Lambda(t)$, the process $\Lambda(S_1), \Lambda(S_2), \ldots$ is a homogeneous Poisson process with rate 1 (called HPP(1) below). By observing only every κth event, it is readily seen that the corresponding process transformed by $\Lambda(\cdot)$ is a gamma renewal process with shape parameter κ. Berman also showed how to generalize to the case when κ is any positive number.

Lakey and Rigdon [22] proposed a special case of Berman's inhomogeneous gamma process, called the modulated power law process, by specializing $\lambda(t)$ to be a power law function. This generalized the much used power law NHPP.

The underlying ideas of Berman [6] were further generalized in [27] to the so-called trend-renewal process (TRP), which can be viewed as a kind of "least common multiple" of the RP and the NHPP and which will be studied in more detail in the sections to come. As will be seen, the simple structure of the TRP makes statistical inference rather attractive. We shall therefore restrict the study of nonparametric statistical inference for intensity-based models to a treatment of the TRP model.

4.1.2 Models for Imperfect Repair Based on Effective Ages

The classical imperfect repair model is the one suggested by Brown and Proschan [8], which assumes that at the time of each failure a perfect repair occurs with probability p and a minimal repair occurs with probability $1 - p$, independently of the previous failure history. We shall call it the BP model for short. A slightly more general model was introduced by Block et al. [7]. In this model, the p of the BP model is generalized to be a function of time t, and upon the nth failure a perfect repair is made with probability $p(a_n)$ and a minimal repair otherwise, where a_n is the age of the system calculated from the last perfect repair (where by convention there is a perfect repair at the beginning of observation). The BP model has in fact been generalized also in several other directions, many of them using the concept of *effective age* to be described below.

Perhaps the most famous extensions of the BP model are the two imperfect repair models suggested by Kijima [21] (see Sect. 4.3.3). Both of these models involve what is called the effective age (or virtual age) of the system. The idea is to distinguish between the system's true age, which is the time elapsed since the system was new, usually at time $t = 0$, and the effective age of the system that describes its present condition when compared to a new system. The effective age is redefined at failures according to the type of repair performed and runs along with the true time between repairs. More precisely, a system with effective age $v \geq 0$ is assumed to behave exactly like a new system that has reached age v without having failed. The hazard rate of a system with effective age v is thus $z_v(t) = z(v + t)$ for $t > 0$, where $z(\cdot)$ is the hazard rate of the time to first failure of a new system. Hence, the distribution function of the time to next failure for a system with effective age v is assumed to be $F_v(t) = F(v + t)/\bar{F}(v)$, where F $(\bar{F})$ is the cumulative distribution function (survival function) corresponding to z.

It should be clear at this stage that models based on effective ages make sense only if the underlying hazard functions $z(\cdot)$ are nonconstant. In fact, if $z(\cdot)$ is constant, then a reduction of effective age would not influence the rate of failures.

Different classes of imperfect repair models can be obtained by specifying properties of the effective age process in addition to the hazard function $z(t)$ of a new system. For this, suppose $a(t)$ is the effective age of the system immediately after the last failure before time t, i.e., at time $S_{N(t-)}$. The effective age at time $t > 0$ is then defined by

$$\mathscr{E}(t) = a(t) + t - S_{N(t-)},$$

which is hence the sum of the effective age after the last failure before t and the time elapsed since the last failure. The process $\mathscr{E}(t)$, named the *effective age process* by Last and Szekli [23], thus increases linearly between failures and may jump (usually *will* jump) at failures. Imperfect repair models can now be defined by stating properties of the repairs via the functions $a(t)$ or $\mathscr{E}(t)$, and these will in general depend on the history of the process.

The simplest examples of effective age processes are for the NHPP and the RP, for which we have, respectively, $\mathscr{E}(t) = t$ and $\mathscr{E}(t) = t - S_{N(t-)}$ (since $a(t)$ is, respectively, $S_{N(t-)}$ and 0 in these cases). It is also seen that the BP model corresponds to letting $\mathscr{E}(t) =$ *time since last perfect repair*.

There is a large literature on reliability modeling using the effective age process. For a recent review, we refer to Hollander et al. [17]. Nonparametric statistical inference in the Brown–Proschan model was first studied by Whitaker and Samaniego [34] and later by Hollander et al. [16]. A more general repair model, containing all the models mentioned above as special cases, was suggested by Dorado et al. [10], who in particular studied nonparametric statistical inference in this model. We shall later in this chapter describe the estimation approach by Whitaker and Samaniego [34] and also the model by Dorado et al. [10].

We shall furthermore review semiparametric estimation in the Peña–Hollander model [29, 30], which in some sense combines the two different approaches of imperfect repair modeling considered in this chapter: via conditional intensities and via effective age processes.

It should be noted that estimation in the last three mentioned models, namely the ones by Whitaker and Samaiego [34], Dorado et al. [10] and Peña et al. [30], need complete knowledge of the effective age processes. This information may be found in repair logs, but is not always available. On the other hand, Doyen and Gaudoin [11] studied classes of effective age models based on deterministic reduction of effective age due to repairs, for which observation of the repair characteristics are not needed.

4.2 The Trend-Renewal Process

The idea behind the trend-renewal process is to generalize the property of the NHPP, described in Sect. 4.1.1 above, namely that $\Lambda(S_1), \Lambda(S_2), \ldots$ is HPP(1) when $S_1, S_2, \ldots$ form an NHPP with cumulative intensity function $\Lambda(\cdot)$.

The trend-renewal process (TRP) is defined simply by allowing this HPP(1) to be any renewal process RP(F), i.e., a renewal process with interarrival distribution F. Thus, in addition to the function $\lambda(t)$, for a TRP, we need to specify the distribution function F of a renewal process. Formally, we can define the process TRP(F, $\lambda(\cdot)$) as follows:

Definition 4.1 *Trend-Renewal Process (TRP)* [27] Let $\lambda(t)$ be a nonnegative function defined for $t \geq 0$, and let $\Lambda(t) = \int_0^t \lambda(u)du$. The process $S_1, S_2, \ldots$ is called TRP(F, $\lambda(\cdot)$) if the transformed process $\Lambda(S_1), \Lambda(S_2), \ldots$ is RP(F), that is if the $\Lambda(S_i) - \Lambda(S_{i-1})$; $i = 1, 2, \ldots$ are i.i.d. with distribution function F. The function $\lambda(\cdot)$ is called the *trend function*, while F is called the *renewal distribution*. In order to have uniqueness of the model specification, it is assumed that F has expected value 1. It is also assumed that $\lim_{t \to \infty} \Lambda(t) = \infty$.

Figure 4.1 illustrates the definition. For an NHPP($\lambda(\cdot)$), i.e., an NHPP with intensity function $\lambda(\cdot)$, the RP(F) would be HPP(1). This shows that the TRP includes the NHPP as a special case by letting F be the standard exponential distribution. Further, if G is a lifetime distribution with finite expectation μ, then with $\lambda(t) = 1/\mu$ and $F(t) = G(\mu t)$, we have RP(G) = TRP(F, $\lambda(\cdot)$). Thus, all RPs with interarrival distributions with finite expectation are TRPs.

To motivate the TRP model as a model for imperfect repair, suppose that failures of a particular system correspond to replacement of a major part, for example, the engine of a tractor [3], while the rest of the system is not maintained. Then, if the rest of the system is not subjected to wear, a renewal process would be

Fig. 4.1 The defining property of the trend-renewal process

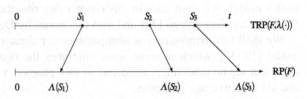

Fig. 4.2 Conditional
intensities for repairable
system with failures observed
at times $t = 1.0$ and $t = 2.25$.
NHPP (*solid*); TRP (*dashed*);
RP (*dotted*)

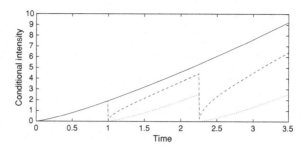

a plausible model for the observed failure process. In the presence of wear, on the
other hand, an increased replacement frequency is to be expected. This is achieved
in a TRP model by accelerating the internal time of the renewal process according
to a time transformation $\Lambda(t) = \int_0^t \lambda(u)du$, which represents the cumulative wear.
The TRP model thus has some similarities to accelerated failure time models.

It can be shown [27] that the *conditional intensity function* (Sect. 3.1.2) for the
TRP$(F, \lambda(\cdot))$ is

$$\gamma(t) = z(\Lambda(t) - \Lambda(S_{N(t-)}))\lambda(t), \tag{4.1}$$

where $z(\cdot)$ is the hazard rate corresponding to F. This is a product of one factor,
$\lambda(t)$, that depends on the age t of the system and one factor that depends on the
time from the last previous failure. However, this time is measured on a scale
depending on the intensity of failures.

Figure 4.2 illustrates the basic difference between the NHPP, RP and TRP as
regards the behavior of conditional intensities. In this example, the three conditional intensities are identical until the first failure, at time $t = 1.0$. After repair,
the intensity for the NHPP starts from the value it had right before the failure, due
to a minimal repair being performed. The lower curve, corresponding to the RP,
represents a perfect repair and resets the conditional intensity at the failure to the
value at $t = 0$ and continues in the same manner as the curve did from time $t = 0$.
The conditional intensity of the TRP, on the other hand, is a product of a factor
corresponding to the underlying renewal distribution, and a trend function which is
here assumed to be increasing in t. This leads to a conditional intensity between
those for the minimal and perfect repair, hence illustrating an effect of imperfect
repair. Similar effects for the three processes are seen at the second failure at
$t = 2.25$.

Figure 4.3 shows the conditional intensity of the TRP in a case where the
renewal distribution has a decreasing failure rate represented by a Weibull
distribution with shape parameter 0.5. This leads to an increase in conditional
intensity after each failure and repair, which for example could be caused by bad
repairs taking place. Thus, new failures will have a tendency to occur right after a
repair. Looking at the figure, the conditional intensity is decreasing for some time
after a repair, but since the trend function is here assumed to be rather strongly
increasing, $\lambda(t) = t^2$, the overall tendency is an increasing intensity of failures.

Fig. 4.3 Conditional intensities for a TRP with failures observed at times $t = 1.0$ and $t = 2.25$. The renewal distribution F is a Weibull distribution with shape parameter 0.5; the trend function is $\lambda(t) = t^2$

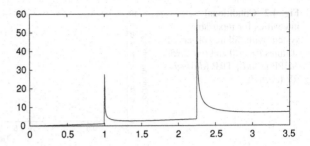

Suppose now that a single system has been observed in $[0, \tau]$, with failures at $S_1, S_2, \ldots, S_{N(\tau)}$. Recall from Andersen et al. [2] that the likelihood function for a counting process with conditional intensity $\gamma(t)$ is given by

$$L = \left\{ \prod_{i=1}^{N(\tau)} \gamma(S_i) \right\} \exp \left\{ - \int_0^\tau \gamma(u) du \right\}. \tag{4.2}$$

The likelihood function of the TRP is now obtained by substituting (4.1) into (4.2), giving

$$L = \left\{ \prod_{i=1}^{N(\tau)} z(\Lambda(S_i) - \Lambda(S_{i-1}))\lambda(S_i) \right\} \exp \left\{ - \sum_{i=1}^{N(\tau)} \int_{S_{i-1}}^{S_i} z(\Lambda(u) - \Lambda(S_{i-1}))\lambda(u) du \right\}$$

$$\times \exp \left\{ - \int_{S_{N(\tau)}}^\tau z(\Lambda(u) - \Lambda(S_{N(\tau)}))\lambda(u) du \right\}.$$

By making the substitution $v = \Lambda(u) - \Lambda(S_{i-1})$, introducing the cumulative hazard $Z(t) = \int_0^t z(v) dv$ and taking the log, we get the log likelihood function

$$l = \log L = \sum_{i=1}^{N(\tau)} \{ \log(z(\Lambda(S_i) - \Lambda(S_{i-1}))) + \log(\lambda(S_i)) - Z(\Lambda(S_i) - \Lambda(S_{i-1})) \}$$
$$- Z(\Lambda(\tau) - \Lambda(S_{N(\tau)})). \tag{4.3}$$

4.2.1 Nonparametric Estimation of the Trend Function of a TRP

Chapters 2 and 3 of this book study in particular nonparametric estimation of the interarrival distribution F in the case of an RP and the intensity function $\lambda(t)$ in the case of an NHPP. Since the TRP by its definition contains both a renewal distribution F and a trend function $\lambda(t)$, either of the two could be estimated nonparametrically, even both of them. We have chosen, however, to concentrate on the estimation problem where $\lambda(t)$ is assumed completely nonparametric, while the renewal distribution $F = F(\cdot; \theta)$ is given on parametric form with hazard rate $z(t; \theta)$ and expected value 1, with θ in some suitable finite dimensional parameter space Θ. In order to cover the NHPP as a special case, we find it convenient to assume that there is a parameter value $\theta^{(0)}$ such that $z(t; \theta^{(0)}) \equiv 1$, i.e., such that $F(\cdot; \theta^{(0)})$ is the standard exponential distribution.

One reason for considering only $\lambda(t)$ in a nonparametric fashion is that it is often the most interesting object of a data study. Furthermore, since the NHPP is a TRP with renewal distribution being the standard exponential distribution, we believe that a TRP with, say, a gamma or Weibull renewal function leads to a case of sufficient interest and usefulness. Indeed, data from repairable systems will often have a tendency to be "almost" like an NHPP, so any relaxation of the standard exponential assumption for the renewal distribution leads to an interesting robustification of the NHPP model.

Note, on the other hand, that if the time trend is believed to be negligible, we have essentially a renewal process and can use methods from Chap. 2 in a possible nonparametric estimation. A more elaborate theory would of course be needed to handle the case when both the trend function and the renewal distribution are nonparametric.

Heggland and Lindqvist [15] considered nonparametric estimation of an assumed monotone trend function $\lambda(\cdot)$ when F is parametric, in particular Weibull-distributed. Later, Lindqvist [26] has considered the case when $\lambda(\cdot)$ is a general nonnegative function. We will review both of these approaches.

A basic assumption to be made here is that there is a single system under consideration. In the algorithms presented in the following, we will maximize the log likelihood (4.3) with respect to the trend function $\lambda(\cdot)$ and the parameter θ of $z(\cdot) \equiv z(\cdot; \theta)$ and $Z(\cdot) \equiv Z(\cdot; \theta)$. This will be done in an iterative manner by alternatively maximizing with respect to $\lambda(\cdot)$ when θ is held fixed, and with respect to θ when $\lambda(\cdot)$ is held fixed. In the latter case, it is seen that the maximization with respect to θ can be done by first computing the transformed interevent times $X_i = \Lambda(S_i) - \Lambda(S_{i-1})$ for $i = 1, 2, \ldots, N(\tau) + 1$, where $S_0 = 0$ and $S_{N(\tau)+1} = \tau$, and then maximizing

$$\sum_{i=1}^{N(\tau)} \{\log z(X_i; \theta) - Z(X_i; \theta)\} - Z(X_{N(\tau)+1}; \theta), \qquad (4.4)$$

which is seen to be the ordinary log likelihood function for maximum likelihood estimation of θ for data transformed by $\Lambda(\cdot)$ (reconsider Fig. 4.1 for the idea of this transformation).

4.2.1.1 The Direct Kernel Smoothing Estimator

From the definition of the TRP(F, $\lambda(\cdot)$), it follows that

$$N^*(u) \equiv N(\Lambda^{-1}(u))$$

is a renewal process with interarrival distribution F. This is simply the process on the lower axis of Fig. 4.1. Letting $M(t) = E(N(t))$ be the mean function of $N(t)$ (Sect. 3.1.2), the mean function of $N^*(u)$ is given by $M^*(u) = M(\Lambda^{-1}(u))$.

Equivalently,

$$M(t) = M^*(\Lambda(t)) \tag{4.5}$$

for all t. A standard asymptotic result for renewal processes gives that

$$\lim_{u \to \infty} M^*(u)/u = 1$$

(since F is here assumed to have expectation 1.) From this, we readily get

$$\lim_{t \to \infty} M(t)/\Lambda(t) = \lim_{t \to \infty} M^*(\Lambda(t))/\Lambda(t) = 1,$$

which can be interpreted to say that for large t one may approximate $\Lambda(t)$ by $M(t)$. We are, however, presently concerned with the estimation of $\lambda(t)$. Differentiating (4.5) we get the rocof

$$m(t) = m^*(\Lambda(t))\lambda(t), \tag{4.6}$$

where $m^*(\cdot)$ is the renewal density corresponding to the renewal distribution F of the TRP. By the renewal density theorem [33], conditions may be specified ensuring that $m^*(u) \to 1$ as $u \to \infty$ (again using that F has expected value 1). Hence, (4.6) indicates that for large t, $m(t)$ and $\lambda(t)$ are approximately equal.

This suggests that $\lambda(t)$, at least for sufficiently large t, can be approximately estimated by an estimate of $m(t)$. A naive estimator for this would be the kernel estimator

$$\hat{m}(t) = \frac{1}{h} \sum_{i=1}^{N(\tau)} w\left(\frac{t - S_i}{h}\right), \tag{4.7}$$

where w is a bounded density function, symmetric around 0 and h is a bandwidth to be chosen. This is the kernel estimator considered in Sect. 3.2.1 in the case of NHPPs. It is also a possible estimator of the rocof in general. To see this, note first that we may approximate w by a convex combination of uniform densities on intervals $[-a, a]$ for $a > 0$. Assume therefore for now that w is such a uniform density. Then, assuming that t is not too close to the endpoints of $[0, \tau]$, we have

$$\hat{m}(t) = \frac{1}{h}\frac{1}{2a} \cdot \#\{i : t - ah < S_i < t + ah\}$$

$$= \frac{1}{2ah}[N(t + ah) - N(t - ah)],$$

so that

$$E(\hat{m}(t)) = \frac{1}{2ah}[M(t+ah) - M(t-ah)]$$

$$\approx \frac{1}{2ah} \cdot m(t) \cdot 2ah = m(t)$$

It is easy to see that the approximation $E(\hat{m}(t)) \approx m(t)$ would hold also for convex combinations of such uniform densities, motivating the approach.

By the connection to renewal processes via (4.5), it is possible, at least in principle, to compute or approximate the expected value and variance of $\hat{m}(t)$. We shall, however, not pursue this here since our aim is to estimate $\lambda(t)$. On the other hand, the approach of the present subsection suggests that the *direct kernel estimator* (4.7) is presumably a good starting function for estimation of $\lambda(\cdot)$ when using the recursive algorithm to be considered in the following.

4.2.1.2 Maximum Likelihood Weighted Kernel Estimation

Consider the expression (4.1) for the conditional intensity function of a TRP. Now $z(\Lambda(t) - \Lambda(S_{N(t-)})) \equiv Y(t)$ is a predictable process, so $\gamma(t) = Y(t)\lambda(t)$ is on the form of Aalen's multiplicative intensity model [2]. For this case, Ramlau-Hansen [31] suggested a kernel estimator of the form

$$\hat{\lambda}(t) = \frac{1}{h}\sum_{i=1}^{N(\tau)} w\left(\frac{t-S_i}{h}\right) Y^{-1}(S_i).$$

As argued by Lindqvist [26], this motivates the use of a weighted kernel estimator for $\lambda(\cdot)$ of the form

$$\lambda(t;a) = \frac{1}{h}\sum_{i=1}^{N(\tau)} w\left(\frac{t-S_i}{h}\right) a_i. \tag{4.8}$$

The clue is then to substitute (4.8) into the log likelihood (4.3) and maximize with respect to the weights $a = (a_i; i = 1, 2, ..., N(\tau))$ in addition to θ. Such an approach is motivated by Jones and Henderson [18, 19] who study ordinary kernel density estimation using variable weights and/or variable locations of the observations.

As indicated earlier, we will maximize (4.3) in an iterative manner, essentially alternating between maximization with respect to θ and the weights $a = (a_i)$ of (4.8). Note again that maximization with respect to θ for given trend function $\lambda(\cdot)$ is the same as maximum likelihood estimation using transformed data (see (4.4)).

In order to simplify the computations, we have found it convenient to modify (4.3) by using the approximation

$$\Lambda(S_i; a) - \Lambda(S_{i-1}; a) \approx \lambda(S_i; a)(S_i - S_{i-1}) \equiv \lambda(S_i; a)T_i \qquad (4.9)$$

for $i = 1, \ldots, N(\tau) + 1$. The corresponding approximation of (4.3) is hence

$$l_a(\theta) = \sum_{i=1}^{N(\tau)} \{\log(z(\lambda(S_i; a)T_i; \theta)) + \log(\lambda(S_i; a)) - Z(\lambda(S_i; a)T_i; \theta)\} \qquad (4.10)$$
$$- Z(\lambda(\tau; a)T_{N(\tau)+1}; \theta),$$

where $T_{N(\tau)+1} = \tau - S_{N(\tau)}$.

The computational algorithm can then be described as follows:

Algorithm for Maximum Likelihood Weighted Kernel Estimation

Let $\theta = \theta^{(0)}$, so that $N(t)$ is an NHPP with intensity $\lambda(\cdot)$. Then compute the direct kernel smoothing estimator

$$\lambda^{(1)}(t) = \frac{1}{h} \sum_{i=1}^{N(\tau)} w\left(\frac{t - S_i}{h}\right). \qquad (4.11)$$

For $m = 1, 2, \ldots$ (until convergence) do:

1. Using the approximation (4.9), compute $X_i^{(m)} = \lambda^{(m)}(S_i)T_i$ for $i = 1, 2, \ldots, N(\tau) + 1$, and let $\theta = \theta^{(m)}$ be the maximizer of (4.4) when $X_i = X_i^{(m)}$.
2. Substitute $\theta = \theta^{(m)}$ in (4.10) and maximize (4.10) with respect to the weights $a = (a_i)$. Let the maximizing a be denoted $a^{(m)}$ and let $\lambda^{(m+1)}(t) = \lambda(t; a^{(m)})$.

The estimates of θ and $\lambda(\cdot)$ are, respectively, the final $\theta^{(m)}$ and $\lambda^{(m)}(t)$.

4.2.1.3 Specializing to a Weibull Renewal Distribution

Note first that a Weibull distribution with shape parameter b and expected value 1 has hazard rate given by

$$z(t; b) = b[\Gamma(b^{-1} + 1)]^b t^{b-1}. \qquad (4.12)$$

If the renewal distribution is as given by (4.12), then the log likelihood (4.10) can be written

$$l_a(b) = N(\tau) \log b + N(\tau)b \log \Gamma(b^{-1} + 1) + \sum_{i=1}^{N(\tau)} \{b \log(\lambda(S_i; a)T_i) - \log T_i$$
$$- [\Gamma(b^{-1} + 1)\lambda(S_i; a)T_i]^b\} - [\Gamma(b^{-1} + 1)\lambda(\tau; a)T_{N(\tau)+1}]^b. \qquad (4.13)$$

In the example below, taken from Lindqvist [26], we use the partial derivatives of l_a in a steepest ascent approach in order to maximize $l_a(b)$ with respect to the a_i for given value of b. The b is then, in alternating steps, estimated by maximum likelihood using (4.4). The kernel used in the example is the Epanechnikov kernel.

Example   U.S.S. Halfbeak Data

Consider again the U.S.S. Halfbeak Diesel Engine data studied in Sect. 3.4.2, but assume now that the failure times follow a TRP with trend function $\lambda(t)$ and a Weibull renewal distribution with shape parameter b. The results for nonparametric estimation of $\lambda(t)$ for three different choices of the bandwidth h are given in Table 4.1 and Fig. 4.4.

As the table shows, there is a tendency for higher maximum value of the likelihood for small values of h. This is reasonable as a small h means following closely the obtained data, with overfitting if h is chosen too small. The fact that the likelihood increases with decreasing h also indicates that in the approach of weighted kernel estimation based on maximizing a likelihood, it is not possible to include h among the parameters in the maximization.

It is also seen from the table that the estimate of b decreases as h increases. The probable reason for this is that a too high degree of smoothing leads to transformed lifetimes $\lambda(S_i)T_i$, which are (modulo a scale factor) closer and closer to the T_i themselves. When the T_i are fitted to a Weibull model, the shape parameter is estimated to 0.63. Thus, increasing h further would reasonably lead to estimated values for b converging to 0.63.

The curves in Fig. 4.4 are much similar to the kernel estimate for the NHPP case given in Fig. 3.5. For both cases, the curves drop from, say, $t = 22$. This may be due to the fact that no edge-correction has been implemented in the present

Table 4.1 Weighted kernel estimation of U.S.S. Halfbeak data

	$h = 2$	$h = 5$	$h = 10$
$\hat{b}$	0.957	0.915	0.870
ℓ_a	−116.91	−120.08	−123.38

Estimates of the shape parameter b and the maximum value of the log likelihood ℓ_a in (4.13)

Fig. 4.4 Weighted kernel estimation of U.S.S. Halfbeak Diesel Engine data. Estimates of $\lambda(\cdot)$ for $h = 2$ (*solid*), $h = 5$ (*dotted*), $h = 10$ (*dot-dash*). (Reprinted from Lindqvist [26].)

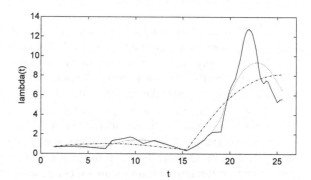

analysis, but a graph of $N(t)$ would also reveal a slight concavity at the right end of the curve.

It is noted by Jones and Henderson [18] in the ordinary density estimation case that the weights a_i obtained at the maximum of the log likelihood are nonzero only for fairly few observations. For the present example, the number of nonzero weights a_i are 20 for $h = 2$, 25 for $h = 5$ and just 9 for $h = 10$, while there are 71 observations. As noted by Jones and Henderson, and as also found in the present computations, the nonzero weights are usually clustered around similar values of the S_i, thus suggesting an even more parsimonius model.

4.2.1.4 Bandwidth Selection Procedures

The maximum likelihood weighted kernel approach considered above involves a bandwidth h that needs to be determined outside the procedure, as indicated above and also discussed by Jones and Henderson [18]. The determination of h for the NHPP case is discussed in Sect. 3.2.2, where a least-squares cross-validation approach is considered. The more general situation considered here is more complicated, however, since the connection to ordinary density estimation is not as clear as for NHPPs. For practical purposes, we suggest that a preliminary bandwidth is chosen by assuming at the first stage of the estimation that the process is an NHPP, and then using the above-mentioned methods from Chap. 3. An adjustment can be done, if needed, at a later stage in order to obtain a reasonable estimated curve for $\lambda(t)$. Indeed, a subjective choice of h seems to be a much used approach in similar situations, at least in cases where an automatic determination of h would not be desirable or necessary.

4.2.2 Nonparametric Estimation of the Trend Function Under Monotonicity Constraints

4.2.2.1 A Monotone Nonparametric Maximum Likelihood Estimator of $\lambda(\cdot)$

Bartozyński et al. [4] derived a nondecreasing NPMLE of $\lambda(\cdot)$ for an NHPP, see Chap. 3. The approach was extended to the TRP case by Heggland and Lindqvist [15]. The following presentation is based on their article.

More specifically, we will in this subchapter seek to maximize the log likelihood function (4.3) under the condition that $\lambda(\cdot)$ belongs to the class of nonnegative, nondecreasing functions on $[0, \tau]$. The case when $\lambda(\cdot)$ is nonincreasing is much similar and will not be considered here [15].

Again, we shall assume that the renewal distribution is parametric, of the form $F = F(\cdot; \theta)$. Observe first that the optimal $\lambda(t)$ must consist of step functions closed on the left with no jumps except at some of the failure time points.

To see this, suppose $\bar{\lambda}t$ is a nondecreasing function that maximizes l in (4.3). Look at the interval $[S_{i-1}, S_i)$ for some fixed i, $1 \le i \le N(\tau) + 1$, where as before we define $S_{N(\tau)+1} = \tau$. Let $v = \int_{S_{i-1}}^{S_i} \bar{\lambda}(u)du$. If we now choose $\lambda(t) \equiv v/(S_i - S_{i-1})$ on $[S_{i-1}, S_i)$, then obviously $\int_{S_{i-1}}^{S_i} \lambda(u)du$ also equals v, leaving all the terms of l unchanged except the term $\log \lambda(S_i)$. But clearly $\lambda(S_{i-1}) \ge \bar{\lambda}(S_{i-1})$ with equality only if $\bar{\lambda}(t) \equiv \lambda(t)$ on $[S_{i-1}, S_i)$. So unless $\bar{\lambda}(t)$ is constant on every interval of the form $[S_{i-1}, S_i)$, it is possible to increase l without violating the nondecreasing property.

Now let for simplicity $n = N(\tau)$, $\lambda_i = \lambda(S_i)$, $i = 0, 1, ..., n$, let $T_i = S_i - S_{i-1}$, $i = 1, 2, ..., n$ and let $T_{n+1} = \tau - S_n$. The problem of maximizing l in (4.3) is then simplified to the problem of maximizing

$$l = \sum_{i=1}^{n} \{\log z(\lambda_{i-1} T_i; \theta) + \log \lambda_i - Z(\lambda_{i-1} T_i; \theta)\} - Z(\lambda_n T_{n+1}; \theta), \quad (4.14)$$

subject to $0 \le \lambda_0 \le \lambda_1 \le \cdots \le \lambda_n$.

The expression l is similar to (4.10), but while (4.10) is just an approximation, (4.14) is the exact function to maximize in the present case.

The resulting estimation problem can now at least in principle be solved by alternatively maximizing with respect to the θ and the λ_i. In the next subsection, we shall see that the maximization with respect to the λ_i has a nice solution using isotonic regression when the renewal distribution is of Weibull type.

4.2.2.2 Specializing to a Weibull Renewal Distribution

We now consider the problem of estimation of a nondecreasing trend function when the renewal distribution is Weibull with hazard function given by (4.12).

Now (4.14) becomes (putting $r = [\Gamma(b^{-1} + 1)]^b$),

$$l_m = n\log b + n\log r$$
$$+ \sum_{i=1}^{n}\{(b-1)\log\lambda_{i-1} + (b-1)\log T_i + \log\lambda_i - r\lambda_{i-1}^b T_i^b\} - r\lambda_n^b T_{n+1}^b$$
$$= n\log b + n\log r + (b-1)\sum_{i=1}^{n}\log T_i + (b-1)\log\lambda_0 - r\lambda_0^b T_1^b \quad (4.15)$$
$$+ \sum_{i=1}^{n-1}\{b\log\lambda_i - r\lambda_i^b T_{i+1}^b\} + \log\lambda_n - r\lambda_n^b T_{n+1}^b.$$

Suppose that the value of b is known. Let $D_i = rT_{i+1}^b$, $i = 0, 1, ..., n$. Let further $C_0 = (b - 1)/b$, $C_n = 1/b$ and $C_i = 1$ for $i = 1, 2, ..., n - 1$. Let $a_i = \lambda_i^b$, $i = 0, 1, ..., n$. Then, the problem of maximizing (4.15) (for given b) simplifies to finding

$$\max_{a_0,a_1,\ldots,a_n} \sum_{i=0}^{n} \{C_i \log a_i - D_i a_i\}$$

subject to $0 \leq a_0 \leq \cdots \leq a_n$. (Note that with the possible exception of C_0, which may be zero or negative, and D_n, which may be zero, all C_i and D_i are positive.)

The solution to this problem is given by

$$a_0 = a_1 = \cdots = a_{k_1} = \min_{0 \leq t \leq n} \frac{\sum_{j=0}^{t} C_j}{\sum_{j=0}^{t} D_j}, \tag{4.16}$$

$$a_{k_1+1} = a_{k_1+2} = \cdots = a_{k_2} = \min_{k_1+1 \leq t \leq n} \frac{\sum_{j=k_1+1}^{t} C_j}{\sum_{j=k_1+1}^{t} D_j}, \tag{4.17}$$

$$\vdots$$

where k_1 is the largest value of t at which the minimum of (4.16) is attained, $k_2 > k_1$ is the largest value of t at which the minimum of (4.17) is attained, etc. Continue this way until $k_l = n$. The problem is in fact a special case of the isotonic regression problem, see for instance Robertson et al. [32], while the method of solution leading to the above is a special case of the "minimum lower sets algorithm", Robertson et al. [32, pp. 24–25].

Some remarks are appropriate at this stage. If $b < 1$, a_0 will attain a negative value. Since we have required $a_0 \geq 0$, the optimum value of a_0 in this case will be 0. This, unfortunately, gives max $l' = \infty$. To see that the estimator obtained is the MLE, impose the artificial constraint $a_0 \geq \delta > 0$ upon the problem. This will cause $a_0 = \delta$, thus ensuring a finite likelihood, but it will not affect any of the other a's if we choose δ small enough. Then, in the end, let $\delta \to 0$. If the system is observed to the time of the nth failure, $\tau = S_n$, then $D_n = 0$ and $a_n = \infty$, also giving an infinite value of the likelihood function. In this case, we may impose $a_n \leq M < \infty$, and let M approach infinity in the end.

The maximization just described is alternating with maximizations with respect to b. Because of the term $(b-1) \log \lambda_0$ in (4.15), which can be made arbitrarily large by choosing both b and λ_0 close to 0, there is a small problem concerning the convergence of the iteration. As long as b is less than 1 and $\lambda_0 = 0$ for every iteration, we may simply ignore the troublesome term, and the iteration will converge, because every time we obtain a new estimate of b (or $\lambda(\cdot)$), this estimate is the MLE conditioned upon the previous estimate of $\lambda(\cdot)$ (or b). Thus with every new iteration, the unconditional log likelihood function (4.15) will increase. As long as b stays above 1 for every iteration, there is no term in (4.15) that may cause problems, and the iteration converges by the same argument as above.

However, if b from one iteration to the next switches from a value below 1 to a value above 1 (or opposite), we should suddenly include (or exclude) the term $(b-1) \log \lambda_0$ in the log likelihood. This could easily lead to a relatively huge change in the estimates, and b might drop below (or rise above) 1 again. And the iteration procedure could start to oscillate between values of b on both sides of 1.

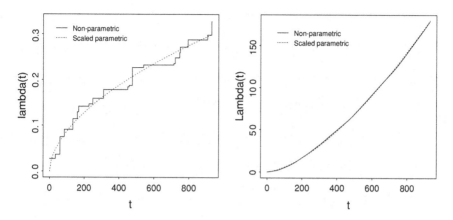

Fig. 4.5 Parametric and nonparametric estimates of $\lambda(t)$ (*left panel*) and $\Lambda(t)$ (*right panel*), a simulated TRP. (Reprinted from Heggland and Lindqvist [15], with permission from Elsevier.)

To avoid this, Heggland and Lindqvist [15] suggested the following modification: Start by assuming that $\lambda_0 = 0$, and ignore the term $(b - 1) \log \lambda_0$. Go through the iteration. If the value of b when it converges is below 1, stop, and use the estimates found. If not, start again, this time including the term $(b - 1) \log \lambda_0$. If the iteration converges, use the estimates now found, if not, use the previous found values.

Example of Monotone Estimation of $\lambda(t)$ A Simulated TRP

Heggland and Lindqvist [15] simulated 200 failure times from a TRP(F, $\lambda(\cdot)$) with $\lambda(t) = 0.01t^{0.5}$ and F being a Weibull distribution with shape parameter $b = 3$. They then used both parametric estimation, assuming $\lambda(t) = \alpha\beta t^{\beta-1}$ and F being Weibull with shape b, and the nonparametric method suggested above, to estimate b and $\lambda(\cdot)$. The purpose of the simulation was to compare the estimates of $\lambda(\cdot)$ and b obtained by the parametric and nonparametric methods, respectively.

The parametric maximum likelihood estimates (with approximate standard deviations in parentheses), computed using the approach of Lindqvist et al. [27], are $\hat{\alpha} = 0.00630(0.00151)$, $\hat{\beta} = 1.550(0.035)$, $\hat{b} = 3.024(0.162)$. On the other hand, the estimate of b under nonparametric estimation of $\lambda(\cdot)$ is 3.269. (It will be shown later how to obtain estimates of bias and standard deviation of this estimate by using bootstrapping.)

The left panel of Fig. 4.5 shows the nonparametric estimate of $\lambda(\cdot)$ together with the parametric estimate $\hat{\lambda}(t) = \hat{\alpha}\hat{\beta} t^{\hat{\beta}-1}$. In addition, the right panel shows the corresponding cumulative functions. As can be seen, there is a good correspondence between the parametric and nonparametric estimates.

Example   U.S.S. Halfbeak Data Reanalyzed

Let us reanalyze the U.S.S. Halfbeak failure data. From the analysis in Sect. 4.2.1 with a nonparametric $\lambda(\cdot)$, it seems fairly reasonable that the system is deteriorating, so that a nondecreasing $\lambda(\cdot)$ is a fair assumption, even if we might suspect

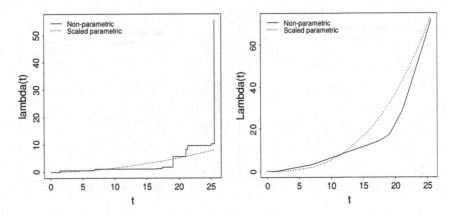

Fig. 4.6 Parametric and nonparametric estimates of $\lambda(t)$ (*left panel*) and $\Lambda(t)$ (*right panel*), U.S.S. Halfbeak data. (Reprinted from Heggland and Lindqvist [15], with permission from Elsevier.)

that $\lambda(t)$ could be decreasing for, say, $t > 22$. A parametric estimation using the power law trend function $\lambda(t) = \alpha\beta t^{\beta-1}$ and a Weibull renewal distribution does however not give a very good fit to this set of data (see below). Still we will use this parametric model for comparison with the nonparametric approach. The parametric MLE's (standard deviations) are: $\hat{\alpha} = 0.00936(0.01225)$, $\hat{\beta} = 2.808(0.402)$, $\hat{b} = 0.762(0.071)$.

The estimate of b under nonparametric estimation of $\lambda(\cdot)$ is 0.937. Plots of both the parametric and the nonparametric estimates of $\lambda(t)$ and $\Lambda(t)$ are found in Fig. 4.6. This time, contrary to the case of the simulated data, the curves based on the parametric estimates are so far from the nonparametrically estimated ones that it is apparently meaningless to compare the different estimates.

The problem is of course that the parametric power law model does not fit the data sufficiently well, as is indicated by the right panel plot of Fig. 4.6. Indeed, a standard confidence interval for b (estimate $\pm 2 \times$ standard deviation) based on the parametric model does not include the value 1, so we reject a null hypothesis of NHPP. On the other hand, the "nonparametric" estimate of b is 0.937 and hence close to 1, which indicates that an NHPP model still is appropriate. This is of course reasonable for this type of data. Likewise is an increasing trend function reasonable and a possible conclusion from this example is that nonparametric estimation of the trend function and parametric estimation of the renewal distribution is a reasonable approach, while a power law trend function does not fit and hence also makes the estimate of b misleading.

4.2.3 Bootstrap Methods for TRPs

Because of the simple structure of the TRP, it is in principle rather straightforward to do bootstrapping in order to estimate bias, standard error and obtain confidence intervals, both for the parameter θ in the renewal distribution and for the trend function $\lambda(\cdot)$. For a general introduction to bootstrapping, we refer to the book by Efron and Tibshirani [12].

The obvious way to generate bootstrap samples from a TRP is to first simulate the estimated renewal process and then transform the arrival times of this renewal process by the inverse mapping of the estimated cumulative trend function $\Lambda(\cdot)$ (see Fig. 4.1). Both nonparametric and parametric bootstrapping are feasible in this way.

The idea is best illustrated by an example. Consider again the estimation for the simulated data considered above, assuming an increasing trend function. The bootstrapping was performed in Heggland and Lindqvist [15] by assuming that the failure times are time-censored at the largest observation which is 931.92. In order to obtain a bootstrap sample, they first drew a sufficiently large number n of observations $Y_1^*, Y_2^*, \ldots, Y_n^*$ from the Weibull cdf $F(x) = 1 - \exp(-x^{3.269})$. Then was used the transformation $S_j^* = \hat{\Lambda}^{-1}(\sum_{i=1}^{j} Y_i^*)$, where $\hat{\Lambda}(\cdot)$ is the nonparametric estimate of $\Lambda(\cdot)$, to obtain a sequence of failure times $S_1^*, S_2^*, \ldots, S_n^*$ of which the ones below 931.92 were used as the first bootstrap sample. This procedure was repeated 50 times, to give 50 bootstrap samples.

For each bootstrap sample, the same estimation procedure as for the original sample was performed, to obtain the bootstrap estimates $\hat{b}_i^*$ and $\hat{\lambda}_i^*(\cdot)$, $i = 1, 2, \ldots, 50$.

The bootstrap estimate of the standard deviation of $\hat{b}$ became 0.209, which might be compared to the normal theory estimate of the standard deviation in the parametric case, which is 0.162. The estimates are reasonably close, and we should also expect the former to be larger because of a less specified model.

The mean of the bootstrap estimates, on the other hand, was 3.578, which could indicate that the estimation is somewhat biased. The bootstrap estimate of bias is thus $3.578 - 3.269 = 0.309$, which is rather large, in fact it is almost a factor 1.5 times the estimated standard deviation. Trying to remove the bias by subtraction [12], we get the bias-corrected estimate for b to be $3.269 - 0.309 = 2.960$. This is very close, both to the "true value" of 3, and the estimated value in the parametric case of 3.024.

The left panel of Fig. 4.7 shows the original estimate of $\Lambda(t)$ along with the estimated curves from the 50 bootstrap samples, giving an indication of the variability of $\hat{\Lambda}(\cdot)$. The figure seems to indicate a relatively low degree of variation, slightly increasing with increasing t. Also, the original estimate seems to be placed neatly in the middle of all the bootstrap estimates, so there is no apparent bias in this case.

In the analysis of the U.S.S. Halfbeak data, Heggland and Lindqvist [15] suggested to *not* draw the Y^*'s from the Weibull distribution $F(x) = 1 - \exp(-x^{0.937})$, but rather from the empirical distribution function

$$\hat{F}(x) = \frac{\text{number of } \hat{Y}_i \le x}{n},$$

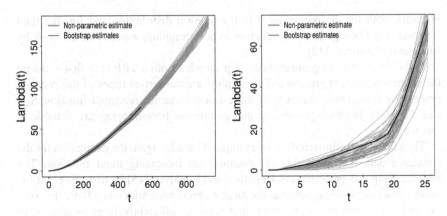

Fig. 4.7 Original nonparametric and bootstrap estimates of $\Lambda(t)$, a simulated TRP (*left panel*) and U.S.S. Halfbeak Diesel Engine data (*right panel*). (Reprinted from Heggland and Lindqvist [15], with permission from Elsevier.)

where $\hat{Y}_i = \hat{\Lambda}(S_i) - \hat{\Lambda}(S_{i-1})$, $i = 1, 2, ..., n$ and $n = 71$ is the total number of failure times. Thus, a *nonparametric bootstrapping* was performed using an empirical renewal distribution instead of an estimated parametric one which was used in the simulated example.

Otherwise, the procedure was exactly the same as for the simulated data example. The estimated standard deviation of $\hat{b}$ is now 0.133, while the mean of the bootstrapped b is 1.016, giving an estimate of the bias, $1.016 - 0.937 = 0.079$. Since this is small compared to the estimated standard deviation, we do not perform a bias correction.

The right panel of Fig. 4.7 shows the original nonparametric estimate $\hat{\Lambda}(\cdot)$ and the bootstrap simulations of $\Lambda(\cdot)$, and in this example, the variability is rather large, particularly in the area where $\hat{\Lambda}(\cdot)$ starts to increase sharply. The variability is larger than in the left panel of the figure. This is probably due to having only 71 failure times, compared to 200 in the left panel, but presumably also due to the fact that we consider a real system where certain changes apparently took place around $t = 19$.

Figure 4.7 also indicates that $\hat{\Lambda}(\cdot)$ for the U.S.S. Halfbeak data perhaps is underestimating the true $\Lambda(\cdot)$, since most of the bootstrap curves apparently are below the NPMLE $\hat{\Lambda}(\cdot)$. To shed light on this, Heggland and Lindqvist [15] computed the bootstrap estimate of standard deviation of $\hat{\Lambda}(t)$ for $t = 19.067$, which is where the sharp increase in $\hat{\Lambda}(t)$ starts. The bootstrap estimate of the standard deviation is 4.286, and the mean of the bootstrap estimates at this point is 16.122. Compared to $\hat{\Lambda}(19.067) = 17.228$, the estimate of bias is $16.122 - 17.228 = -1.116$ which is however relatively small compared to the estimated standard deviation. It was concluded that the apparent underestimation could be due to forcing $\hat{\lambda}(\cdot)$ to be nondecreasing for all t, while this is not necessarily the case for $t > 20$ for these data.

4.3 Models Based on Effective Ages

4.3.1 Nonparametric Statistical Inference for the Brown–Proschan Model

The BP model [8] has already been described in the introduction to this chapter. The basic property is that, following any system failure, a perfect repair or a minimal repair is made with probability p and $1 - p$, respectively, where $0 \leq p \leq 1$. In the present section, we shall review an approach for nonparametric inference in this model presented by Whitaker and Samaniego [34]. If we let F denote the distribution of time to first failure for a new system, then the authors more precisely addressed the problem of estimating the unknown parameters (p, F) using a nonparametric maximum likelihood estimation (NPMLE) approach. First, they concluded that the BP model is not identifiable on the basis of observed interfailure times alone, so that data on the mode of repair after each observed failure would be needed (or imputed in some justifiable fashion) in order to estimate p and F using classical statistical methods.

They therefore consider observations of the form (T_i, Z_i), $i = 1, 2, \ldots$ where the T_i are the interfailure times and the Z_i are indicators of repair, which equal 1 for a perfect repair and 0 for a minimal repair. They furthermore consider the inference problem under an inverse sampling scheme, where failures are recorded until the occurrence of the mth perfect repair.

Let n be the total number of repairs in the available sample, where m of them are perfect repairs and the remaining $n - m$ are imperfect repairs. If F is absolutely continuous with density f, the likelihood associated with the observed interfailure times $T_i = t_i$ and corresponding modes of repair $Z_i = z_i$ ($i = 1, \ldots, n$), is given by

$$L(p, F) = p^m (1 - p)^{n-m} f(x_1) \frac{f(x_2)}{S(x_1)^{1-z_1}} \cdots \frac{f(x_n)}{S(x_{n-1})^{1-z_{n-1}}},$$

where $S = 1 - F$ and x_i is the age of the system just prior to the ith failure. (It should be noted that the x_i can be obtained from the data (t_i, z_i).)

Whitaker and Samaniego [34] then show that the NPMLE of (p, F) is given by $\hat{p} = m/n$ and

$$\hat{F}(t) = \begin{cases} 0 & \text{for } t < x_{(1)} \\ 1 - \prod_{j=1}^{i} k_j/(k_j + 1) & \text{for } x_{(i)} \leq t < x_{(i+1)}, \\ 1 & \text{for } t \geq x_{(n)} \end{cases}$$

where $x_{(1)} < \cdots < x_{(n)}$ are the ordered ages (x_i), $k_i = \sum_{j=i}^{n-1} z_{[j]}$, and $z_{[1]}, \ldots, z_{[n]}$ are the concomitants of $z_1, \ldots, z_n$ corresponding to the ordered values $x_{(i)}$. The authors remark that $z_{[n]}$ necessarily must equal 1 and show that the above estimator is a

NPMLE if $z_{[n-1]} = 1$ and is otherwise a *neighborhood estimator* in the sense of Kiefer and Wolfowitz [20].

Whitaker and Samaniego [34] also prove weak convergence of the process $\sqrt{m}(\hat{F}(t) - F(t))$ to a zero-mean Gaussian process on any fixed interval $[0, T]$ with $F(T) < 1$. Finally, they indicate how the estimation procedure can be extended to the age-dependent imperfect repair model of Block et al. [7].

The large sample results of Whitaker and Samaniego [34] were extended by Hollander et al. [16] using counting process techniques. They furthermore considered the more general situation where m independent systems are observed, and they also obtained a nonparametric asymptotic simultaneous confidence band for the distribution function F.

Example Norsk Hydro Compressor Data [13]

Table 4.2 shows a subset of failure and repair data from a gas compressor system at an ammonia plant of Norsk Hydro [13]. Subsequent intervals between failures T_i (days), the corresponding repair characteristic Z_i, as well as the time since last perfect repair X_i, are recorded for $n = 26$ failures in the period 1969–1975. Here time $t = 0$ corresponds to a specific perfect repair time in the full data set. The original data also report downtime after each failure, and we have for illustration let a *perfect repair* correspond to a downtime of at least 10 h. This has led to $m = 9$ perfect repairs, so $\hat{p} = 9/26 = 0.346$.

Figure 4.8 shows the nonparametric estimate $\hat{S} = 1 - \hat{F}$, plotted together with the estimated survival function $\hat{G}$ obtained under the assumption that all intervals between failures are i.i.d. exponential. Although the difference is not as pronounced as in the example provided by Whitaker and Samaniego [34], it is seen that their estimator, which uses the information about perfect repair, gives more optimistic estimates. This is because low interfailure times T_i ideally should correspond to high ages X_i.

Table 4.2 Subsequent intervals between failures of a gas compressor system at an ammonia plant of Norsk Hydro [13]

i	Z_i	T_i	X_i	i	Z_i	T_i	X_i
1	0	7	7	14	0	42	42
2	1	90	97	15	0	19	61
3	0	66	66	16	0	134	195
4	1	35	101	17	0	29	224
5	0	61	61	18	0	91	315
6	0	32	93	19	1	86	401
7	1	176	269	20	0	132	132
8	1	140	140	21	0	3	135
9	0	20	20	22	1	7	142
10	0	113	133	23	0	13	13
11	0	271	404	24	0	32	45
12	0	49	453	25	1	59	104
13	1	14	467	26	1	180	180

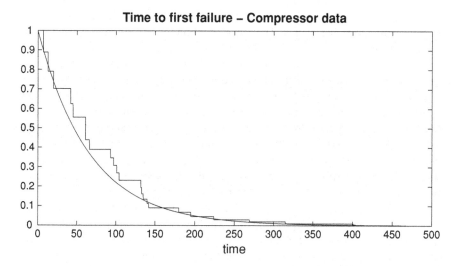

Fig. 4.8 The nonparametric MLE $\hat{S} = 1 - \hat{F}$ and the exponential MLE $\hat{G}$

4.3.2 Nonparametric Statistical Inference for the Peña-Hollander Model

Peña and Hollander [29] proposed a general class of models, which incorporates several aspects of the behavior of recurrent events. Since it involves covariates, it might be considered as a hazard regression model (such models are studied in Chap. 7). However, since the effective age process is a main ingredient of the modeling, we shall present the model in this chapter as an imperfect repair model.

Thus, consider a system being monitored over a time period that is either a fixed or random interval. Associated with the unit is a, possibly time-dependent, covariate vector $x = \{x(s): s \geq 0\}$. Typically, the covariate vector will be related to some environmental or operating condition and may influence the failure performance of the system. Following failures, the system is repaired or replaced, at a possibly varying degree. It is thus assumed some form of imperfect repair. There is furthermore included a possibility for the times between failures to be correlated, possibly because of unobserved *random effects* or so-called *frailties*.

The model of Peña and Hollander [29] seeks to combine all these features and is defined by the following conditional intensity function, conditional on the history of the process up to but not including time t and on the unobserved frailty variable Z:

$$\gamma(t) = Z \lambda_0(\mathscr{E}(t)) \, \rho(N(t-); \alpha) \, g(x(t); \beta). \tag{4.18}$$

The frailty Z is a positive random variable, commonly taken to be gamma-distributed with expected value 1, while $\lambda_0(\cdot)$ is a baseline hazard rate function. Typical choices for the functions ρ and g are $\rho(k; \alpha) = \alpha^k$ and $g(x(t); \beta) = \exp\{\beta x(t)\}$, respectively. Thus, if $\alpha > 1$, the effect of accumulating event occurrences is to accelerate event occurrences, whereas if $\alpha < 1$, the event occurrences decelerate future events. The latter situation is appropriate for example in software debugging.

A special feature of the model is that the effective age process $\mathscr{E}(t)$ appears as argument in the baseline hazard rate function. Effective age processes may occur in many forms as already discussed in the introduction to the present chapter, and the idea is that $\mathscr{E}(t)$ should be determined in a dynamic fashion in conjunction with interventions and repairs that are performed. As considered earlier, the most common forms are $\mathscr{E}(t) = t$ in the case of minimal repairs and $\mathscr{E}(t) = t - S_{N(t-)}$ in the case of perfect repairs. The Peña–Hollander model can however in principle work with any effective age process, for example the process connected to the BP model, where $\mathscr{E}(t)$ is the time since the last perfect repair.

The frailty Z may well be viewed as being the effect of an *un*observed covariate. Systems with a large value of Z will have a larger failure proneness than systems with a low value of Z. Intuitively, the variation in the Z between systems implies that the variation in observed number of failures among the systems is larger than would be expected if the failure processes were identically distributed. Now, since Z is unobservable, one needs to take the expectation of the likelihood that results from (4.18) with respect to the distribution of Z in order to have a likelihood function for the observed data. It is the unconditioning with respect to Z that makes the times between events correlated.

Peña et al. [30] study semiparametric inference in the model (4.18), where $\lambda_0(\cdot)$ is assumed to belong to the general nonparametric class of nonnegative functions, while it is assumed that $\rho(k; \alpha) = \alpha^k$, $g(x(t); \beta) = \exp\{\beta x(t)\}$ and that Z is gamma-distributed with mean 1 and variance $1/\xi$. The unknown parameters are thus (α, β, ξ) in addition to $\lambda_0(\cdot)$. It is furthermore assumed that n independent systems are observed over possibly different time intervals. More precisely, the observations can be written

$$\{[(X_i(v), N_i(v), \mathscr{E}(v)), 0 \le v \le \tau_i], \quad i = 1, 2, \ldots, n\}.$$

Here, the observation lengths τ_i may be random but are usually assumed to be independent of the failure processes and noninformative with respect to parameters of the failure model. A difficulty encountered in estimating $\lambda_0(\cdot)$ is that in (4.18) the argument of $\lambda_0(\cdot)$ is the effective age $\mathscr{E}(t)$ and not t. As explained by Peña [28], this poses difficulties especially in establishing asymptotic properties, because the usual martingale approach [2] does not directly carry through.

Peña et al. [30] are still able to derive, first for the case with no frailties, a profile likelihood for α, β similar to the Cox partial likelihood, which by maximization leads to estimates $\hat{\alpha}, \hat{\beta}$. The baseline $\Lambda_0(\cdot)$ can subsequently be estimated by a Breslow-type estimator. It is suggested to estimate standard errors using the approximate inverse of the partial likelihood information matrix. With the inclusion of frailties, it is furthermore suggested to estimate the parameters α, β, ξ by using an EM algorithm and to use jackknifing to estimate the standard errors. The estimation procedures just described are implemented in the R-package 'gcmrec' [14].

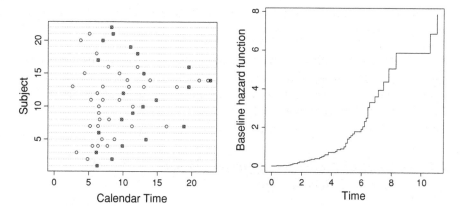

Fig. 4.9 Event plot (*left panel*) and estimated cumulative baseline hazard function (*right panel*) for tractor engine data from Barlow and Davis [3]. Time unit is 10^3 h

Example Tractor Engine Replacements [3]

Barlow and Davis [3] presented a set of engine replacement times (in operating hours) for 22 tractors. An event plot is shown in the left panel of Fig. 4.9. There is a total number of 59 failures, where the number of failures for each system varies from 1 to 6. Lindqvist et al. [27] fitted parametric TRP models to these data and found in particular that neither an RP nor an NHPP model was appropriate models. Since they also found that there is no significant frailty effect, they ended up suggesting a TRP model with increasing trend function as well as a renewal distribution with increasing failure rate. This suggests to model the data by the Peña–Hollander model without frailties [i.e., assuming $Z = 1$ with probability 1 in (4.18)]. Further, the effective age function $\mathscr{E}(t)$ should be chosen to be of the perfect repair type, because as already mentioned the analysis of Lindqvist et al. [27] strongly indicated non-exponentiality of times between events. Thus we shall fit, using the above mentioned R-package 'gcmrec', the model with conditional intensity

$$\gamma(t) = \lambda_0(t - S_{N(t-)})\alpha^{N(t-)}e^{\beta x}, \qquad (4.19)$$

where we let the covariate x be the censoring time for each engine. (Note that the R-package does not allow an estimation without covariates). The motivation for this covariate is that there is no information on how the data are sampled. Thus, it might be of interest to check whether the observation periods for each engine are affected by the frequency of replacements.

The reported estimates of the parameters α and β are, respectively (with standard errors in parentheses), $\hat{\alpha} = 1.85$ (0.247) and $\hat{\beta} = -0.0691$ (0.0338). There is thus an apparent increasing trend in the occurrence of engine replacements, which supports the conclusion of Lindqvist et al. [27]. Further, there is a small (though significant) tendency of lower replacement frequency for engines with long observation time. (This last feature may of course also be a sign of lack of model fit for a model without covariates). An attempt to fit the model *with* frailties gave an estimated inverse variance ξ to be 2.59×10^9, which indeed is a strong indication of no frailty effects. As already noted, this was in fact also the conclusion of Lindqvist et al. [27].

The right panel of Fig. 4.9 shows the estimated cumulative baseline hazard function $\hat{\Lambda}_0(\cdot)$. Note that the convex shape (at least for $t < 8 \times 10^3$ h) indicates an increasing failure rate of the time to failure for a new engine. This is also in accordance with Lindqvist et al. [27].

4.3.3 A General Model for Imperfect Repair Based on Effective Ages

Dorado et al. [10] suggested a general repair model that contains a large class of imperfect repair models as special cases. The definition in their paper is given in terms of conditional distributions of times between failures, T_i. Following Lindqvist [24], we shall instead give an equivalent definition in terms of conditional intensities as follows:

$$\gamma(t) = \theta(t)z\big(a(t) + \theta(t)(t - S_{N(t-)})\big).$$

Here, $z(\cdot)$ is the hazard rate function corresponding to the distribution function $F(\cdot)$ for the time to first failure of the system. In addition, the model is defined in terms of two random processes $a(t)$ and $\theta(t)$. Here, $a(t) = $ effective age immediately after repair of the last failure before t, i.e., at time $S_{N(t-)}$, while $\theta(t) = $ repair characteristic of last repair before t (called "life supplement"), $0 < \theta(t) \le 1$. It should be noted that both $a(t)$ and $\theta(t)$ change their values at failures only and are constant between failures.

It is easy to see that an RP corresponds to $\theta(t) \equiv 1$, $a(t) \equiv 0$, while an NHPP has $\theta(t) \equiv 1$, $a(t) = S_{N(t-)}$.

The two models for imperfect repair suggested by Kijima [21] also turn out to be special cases of the above model. In fact, Kijima's Model I has $\theta(t) \equiv 1$, while the effective age $a(t)$ increases by the amount $D_j T_j$ at the jth failure time S_j, where $D_1, D_2, \dots$ is a sequence of random variables independently distributed on $[0, 1]$ and independent of other processes.

In Kijima's Model II, we have $\theta(t) \equiv 1$, while the effective age $a(t)$ immediately after the jth failure is set to $\sum_{k=1}^{j}(\prod_{i=k}^{j} D_i)T_k$. Equivalently, the effective age after the jth failure equals D_j multiplied by the effective age just before the jth failure.

When D_j is 1 with probability p and 0 with probability $1 - p$, we obtain the BP model. Moreover, the model A by Bathe and Franz [5] is given by letting D_j be 1, c, 0 with respective probabilities p, q, $1 - p - q$. Here $0 < c < 1$ corresponds to an intermediate repair action.

Dorado et al. [10] considered statistical inference in the general repair model when the effective ages $a(t)$ and life supplements $\theta(t)$ are observed at each failure, in addition to the failure times S_i. Using the general counting process framework, they derived estimates and confidence bands for the distribution function F of the time to first failure, under the condition that this has an increasing failure rate.

In real applications, however, exact information on the type of repair and more specifically effective ages and life supplements, are rarely available. Doyen and Gaudoin [11] therefore studied classes of virtual age models based on deterministic reduction of effective age due to repairs, and hence not requiring the observation of repair characteristics. Examples of such models are obtained simply by letting the D_i in Kijima's models (as described above) be replaced by parametric functions. A simple case considered by Doyen and Gaudoin [11] puts $1 - \rho$ for the D_i, where $0 < \rho < 1$ is now a so-called age reduction factor.

References

1. Aalen OO, Borgan O, Gjessing HK (2008) Survival and event history analysis. A process point of view. Springer, New York
2. Andersen PK, Borgan O, Gill RD, Keiding N (1993) Statistical models based on counting processes. Springer, New York
3. Barlow RE, Davis B (1977) Analysis of time between failures of repairable components. In: Fussel JB, Burdick GR (eds) Nuclear systems reliability engineering and risk assessment. SIAM, Philadelphia, pp 543–561
4. Bartozyński R, Brown BW, McBride CM, Thompson JR (1981) Some nonparametric techniques for estimating the intensity function of a cancer related nonstationary Poisson process. Ann Stat 9(5):1050–1060
5. Bathe F, Franz J (1996) Modelling of repairable systems with various degrees of repair. Metrika 43:149–164
6. Berman M (1981) Inhomogeneous and modulated gamma processes. Biometrika 68:143–152
7. Block H, Borges W, Savits T (1985) Age dependent minimal repair. J Appl Probab 22: 370–385
8. Brown M, Proschan F (1983) Imperfect repair. J Appl Probab 20:851–859
9. Cox DR (1972) The statistical analysis of dependencies in point processes. In: Lewis PA (ed) Stochastic point processes. Wiley, New York, pp 55–66
10. Dorado C, Hollander M, Sethuraman J (1997) Nonparametric estimation for a general repair model. Ann Stat 25:1140–1160
11. Doyen L, Gaudoin O (2004) Classes of imperfect repair models based on reduction of failure intensity or virtual age. Reliab Eng Syst Saf 84:45–56
12. Efron B, Tibshirani R (1993) An introduction to the bootstrap. Chapman and Hall/CRC, Boca Raton
13. Erlingsen SE (1989) Using reliability data for optimizing maintenance. Unpublished Masters thesis, Norwegian Institute of Technology, Trondheim, Norway
14. González JR, Slate EH, Peña EA (2009) Package 'gcmrec'. General class of models for recurrent event data. R package version 1.0-3. http://cran.r-project.org/web/packages/gcmrec/
15. Heggland K, Lindqvist BH (2007) A non-parametric monotone maximum likelihood estimator of time trend for repairable systems data. Reliab Eng Syst Saf 92:575–584
16. Hollander M, Presnell B, Sethuraman J (1992) Nonparametric methods for imperfect repair models. Ann Stat 20:879–896
17. Hollander M, Samaniego FJ, Sethuraman J (2007) Imperfect repair. In: Ruggeri F, Kenett R, Faltin FW (eds) Encyclopedia of statistics in quality and reliability. Wiley, New York
18. Jones MC, Henderson DA (2005) Maximum likelihood kernel density estimation. Technical report 01/05. Department of Statistics, The Open University, United Kingdom

19. Jones MC, Henderson DA (2009) Maximum likelihood kernel density estimation: on the potential of convolution sieves. Comput Stat Data Anal 53:3726–3733
20. Kiefer J, Wolfowitz J (1956) Consistency of the maximum likelihood estimator in the presence of infinitely many incidental parameters. Ann Math Stat 27:887–906
21. Kijima M (1989) Some results for repairable systems with general repair. J Appl Probab 26:89–102
22. Lakey MJ, Rigdon SE (1992) The modulated power law process. In: Proceedings of the 45th Annual Quality Congress, pp 559–563
23. Last G, Szekli R (1998) Stochastic comparison of repairable systems by coupling. J Appl Probab 35:348–370
24. Lindqvist B (1999) Repairable systems with general repair. In: Schueller G, Kafka P (eds) Safety and reliability. Proceedings of the European conference on safety and reliability, ESREL '99, Munich, 13–17 Sep 1999. Balkema, Boston, pp 43–48
25. Lindqvist BH (2006) On the statistical modelling and analysis of repairable systems. Stat Sci 21:532–551
26. Lindqvist BH (2010) Nonparametric estimation of time trend for repairable systems data. In: Balakrishnan N, Nikulin M, Rykov V (eds) Mathematical and statistical methods in reliability. Applications to medicine, finance, and quality control. Birkhauser, Boston, pp 277–288
27. Lindqvist BH, Elvebakk G, Heggland K (2003) The trend-renewal process for statistical analysis of repairable systems. Technometrics 45:31–44
28. Peña EA (2006) Dynamic modeling and statistical analysis of event times. Stat Sci 21: 487–500
29. Peña E, Hollander M (2004) Models for recurrent events in reliability and survival analysis. In: Soyer R, Mazzuchi T, Singpurwalla N (eds) Mathematical reliability: an expository perspective. Kluwer, Boston, pp 105–123
30. Peña EA, Slate EH, González JR (2007) Semiparametric inference for a general class of models for recurrent events. J Stat Plan Inference 137:1727–1747
31. Ramlau-Hansen H (1983) Smoothing counting process intensities by means of kernel functions. Ann Stat 11:453–466
32. Robertson T, Wright FT, Dykstra RL (1988) Order restricted statistical inference. Wiley, New York
33. Smith WL (1958) Renewal theory and its ramifications. J R Stat Soc Ser B 20:243–302
34. Whitaker LR, Samaniego FJ (1989) Estimating the reliability of systems subject to imperfect repair. J Am Stat Assoc 84:301–309

Part III
White-Box Approach:
The Physical Environment

Finally, we have models that are as close as possible to a full description of the real system. In this situation, we consider that a deep understanding of the different mechanisms of failure is required and therefore building a physic-based model is needed. The complexity and the computing burden implicit in the model increase with the degree of mathching between the model and the physic of the system, so a pure white-box model formulation is not feasible in practice. For this reason, we will consider white-box models only as far as the implicit mathematical background is tractable. In particular, we consider the white-box approach from three different points of view.

- On the one hand, the first step in determining the performance conditions of a reliability system is to examine its current state of functioning (failure). The procedure consists of splitting the global structure that represents the system into smaller units referred to as components. The number of components in the system will determine the complexity of the model as well as its accuracy in describing the physical system. A linking function that expresses the state of the system given the state of each component is desirable. The easiest formulation allows only two performance states for the system and for each component, the functioning one and the failed one. However, to be more realistic, one needs to realise that most systems run at different levels of performance. As a consequence, the state space is generalised to be a discrete or even a continuous set. Obtaining an expression for the structure function of a continuum system becomes difficult. A procedure to build a structure function for a continuum system by using multivariate nonparametric regression techniques, in which certain analytical restrictions on the variable of interest must be considered, is applied.
- If we take into account the time evolution of the system, a multistate model can be formulated within the scope of a white-box approach. The reason is as follows. In a multistate model we define a state space describing the

stochastic behaviour of the system. The states are defined by considering the intrinsic physical behaviour of the system, that is, the transition from one state to another in the model means, physically speaking, the failure or the repair of one of the components of the system. The most popular multistate models are Markov processes. In practice, the Markovian approach may not always hold, since the hypothesis that the sojourn times in states are exponentially distributed is not admissible. So a more general formulation is necessary and therefore we consider semi-Markov processes as models for the behaviour of a system that evolves in time, passing through a certain state space. The main characteristics of the process, such as transition rate, availability and ROCOF, among others, are estimated by using nonparametric procedures.

- Finally, the physical description of the deterioration process of a system may rely on considering several (exogenous and endogenous) factors that are commonly referred to as covariates. The inclusion of this kind of information in the deterioration model may be attempted in several ways, which leads us to study different regression models for the lifetime. There is a vast literature on semi-parametric models that addresses the relationship between lifetime and covariates. Maybe the best known of all is the extensively used Cox proportional hazard model, which supposes proportionality of the hazard rates of two items defined by different sets of covariates. This assumption may, in many cases, be very restrictive. An appropriate alternative could be supplied by nonparametric hazard models. The nonparametric estimation of the hazard rate, given a vector of covariates, may be tackled in several ways. One approach is by smoothing a conditional (given a vector of covariates) Nelson–Aalen estimator. Other works recently published, define an estimator of the conditional hazard rate as the ratio of nonparametric estimators of the conditional density and the survivor function.

Chapter 5
Systems with Multi-Components

5.1 From Binary to Continuous-State Reliability Systems

Technical systems are designed to perform an intended task with an admissible range of efficiency. According to this idea, it is permissible that the system runs among different levels of performance, in addition to the failed one and the functioning one. As a consequence, reliability theory has evolved from binary-state systems to the most general case of continuous-state systems, in which the state of the system changes over time through some interval on the real number line. In this context, obtaining an expression for the structure function becomes difficult, compared to the discrete case, with difficulty increasing as the number of components of the system increases. In this chapter, we propose a method to build a structure function for a continuum system by using multivariate nonparametric regression techniques, in which certain analytical restrictions on the variable of interest must be taken into account. In other words, given a set of observed values of the system and component states, the aim is to obtain an estimate of the structure function. To do this, we propose applying a procedure based on multivariate smoothing and isotonic regression methods, adapted to the particular characteristics of the problem that we consider here.

5.1.1 Introduction

Most research in reliability models have traditionally concentrated on a binary formulation of systems behavior, that is, models allow only two states of functioning for a system and its components: perfect functioning and complete failure. However, in practice, many systems may experience continuous degradation so that they can exhibit different levels of performance between the two extreme cases of full functioning and fatal failure. A typical example is a system subject to

M. L. Gámiz et al., *Applied Nonparametric Statistics in Reliability*,
Springer Series in Reliability Engineering, DOI: 10.1007/978-0-85729-118-9_5,
© Springer-Verlag London Limited 2011

wear, which degrades continuously with time, so its performance properties decrease progressively and, as a consequence, it is necessary to consider a wider specification of the state space in order to have a more precise and appropriate description of the behavior of the system at each time.

Baxter [4, 5] first introduced continuum models for reliability systems, and, since then, a wide variety of performance measures have been defined and calculated to be valid for binary, multi-state and continuum systems (see Brunelle and Kapur [8], for an extensive review and more recently, [21]). In particular, the structure function of the system, which represents the link function between system state and its components, has been a subject of primary interest in the field of reliability engineering. Since the reliability evaluation can be a very difficult problem in practice, even for relatively simple systems (see Pourret et al. [25] and Lisnianski [20], for instance), it seems reasonable that to have a procedure for modelling the relationship between the system state and its components may assist efficiently in the reliability assessment of complex systems.

For binary systems, the structure function can be determined when either the minimal paths or minimal cuts are known [3]. If it is the case of more than two states, several procedures have been developed in order to generalize the concept of binary coherent structure to a multi-state configuration, and so, the structure function can be specified via a finite set of boundary points, see El Neweihi et al. [11] and Natvig [24] for a complete treatment of the problem. Later, Aven [1] justifies the introduction of multi-state models by the needs in some areas of application, such as gas/oil production and transportation systems, where a binary approach would give a poor representation of the real situation. He investigates the problem of computing relevance performance measures for a multi-state monotone system, some comparisons of the accuracy of such computations and the ones obtained by Monte Carlo methods are presented.

In Pourret et al. [25], a Boolean model is derived in order to describe the state of a multi-state system, revealing that the reliability evaluation of a system is a difficult task from a practical viewpoint, even for systems not excessively complexes. Meng [22] carries out a comparative study of two upper bounds for the reliability of a multi-state system, generalizing the binary case.

In case of a continuous system, if the structure function cannot be determined based on qualitative characteristics (for instance, series or parallel structures) or boundary point analysis, approximation methods are required. To that effect, several treatments of the problem have been carried out. Levitin [18] investigates an approach based on the universal generating function technique. The method consists of a discrete approximation of the continuous-state system performance by using a finite multi-state system and the purpose is to construct upper and lower bounds for the reliability measures of the continuous system.

Given the high difficulty inherent in the analytical evaluation of the performance of a continuous system, a new approach based on empirical methods has been introduced recently. Brunelle and Kapur [7] proposed a multivariate interpolation procedure by which a structure function for a continuous system is built starting from a scatterplot that represents the value of the state of the system at some values of the components states.

Under this empirical perspective, Gámiz and Martínez-Miranda [13] propose a new technique that assumes a regression model for the structure function of a continuous system. The main purpose is to construct a structure function for the system given an observed set of states of the system and its components. The main idea is based on the use of a class of monotone nonparametric regression techniques. When the regression function is monotone in one or more explanatory variables, it is appropriate to use any monotonic regression technique. Several numerical procedures are available from the specialized literature. Pool Adjacent Violators Algorithm (*PAVA*) is the method of most widespread use. It was first proposed to solve the one-dimensional case, when only one explanatory variable is considered, and it has been subsequently generalized to higher dimension problems (see Burdakow et al. [9]). Other solutions have been proposed to the problem of monotonic regression in one or more dimensions; Hall and Hwang [15], Mukarjee and Stem [23], for instance. However, *PAVA* can be easily implemented and extended to the case of multivariate regression. Moreover, it can be applied after a nonparametric estimation method has been applied to a dataset.

Hence, the problem treated in this chapter can be summarized as follows: find a nonparametric monotone surface that fits properly a given dataset. With this aim, a solution that combines local smoothing and isotonic regression is explored.

5.1.2 The Structure Function

Let V be the random variable that denotes the state of a system composed by r elements, which are assumed to be mutually independent. The state of component i is a random variable U_i, for any $i = 1, 2, \ldots, r$. The structure function captures the relationships between the components of a system and the system itself, in such a way that the state of the system is known from the states of its components through the structure function.

Definition 5.1 (*Structure Function*) Let $\mathbf{U} = (U_1, U_2, \ldots, U_r)$ be a random row-vector that specifies the states of the components of a reliability system. Let $\mathbf{u} = (u_1, u_2, \ldots, u_r)$ denote a particular value of $\mathbf{U}$. The structure function of the system is defined as a function φ that expresses the state of the system in terms of the states of its components, in symbols, $\varphi(u_1, u_2, \ldots, u_r) = v$, where v is a particular value of the random variable V, defined above.

We consider here *coherent reliability systems*, which implies that some restrictions must be imposed to the function φ. In words, on the one side the system is monotone, which means that the improvement of any component does not degrade the state of the system. Moreover, when all components occupy the minimal (maximal) state, then the system occupies its minimal (maximal) state.

And finally, there are no *irrelevant* components in the system. These assertions may be expressed as in the following definition, where by convenience, which does not impose any loss of generality, we assign for each component as well as for the system the minimal state equal to 0 and likewise the maximal state is assumed to be equal to 1.

Definition 5.2 (*Coherent System*) A reliability system is said to be coherent if the following conditions are satisfied:

- *Monotonicity*
  For fixed values $u_1, \ldots, u_{i-1}, u_{i+1}, \ldots, u_r$,

$$\varphi\left(u_1, \ldots, u_{i-1}, v^1, u_{i+1} \ldots, u_r\right) \leq \varphi\left(u_1, \ldots, u_{i-1}, v^2, u_{i+1} \ldots, u_r\right),$$

  for any $v^1 \leq v^2$.
- *Proper extrema*

$$\varphi(0, 0, \ldots, 0) = 0 \text{ and } \varphi(1, 1, \ldots, 1) = 1$$

- *Component relevance*

$$\sup\{\varphi(\mathbf{u}|1_i) - \varphi(\mathbf{u}|0_i); \mathbf{u} \in S\} > 0,$$

for all $i = 1, 2, \ldots, r$. Where the notation $(\mathbf{u}|v_{i.}) = (u_1, \ldots, u_{i-1}, v, u_{i+1}, \ldots, u_r)$, for $v \in [0, 1]$ is used.

According to the level of description one wishes for representing the behavior of reliability systems, several approaches have been considered in the literature. The situation may be summed up in the following classification.

Binary Structures. Under this approach, the behavior of the system is modeled assuming that only two states are distinguished for the system and for each component. That is, the functioning state, which is usually represented by the value of 1, and the failure state, represented as 0. So, V and U_i, for each $i = 1, 2, \ldots, r$, are considered to be binary random variables with values $\{0, 1\}$. In this case, the structure function of the system is a mapping $\varphi : \{0, 1\}^r \mapsto \{0, 1\}$, which is non-decreasing in each argument.

Discrete Structures. In this paragraph, we refer to models that expand the dichotomy of the binary approach to allow distinguishing between several levels of failure and functioning. In the most general description, we limit ourselves to consider that a finite number of non-negative numbers are sufficient to characterize the stochastic behavior of components and system. So, for a particular component we assume that for any $i = 1, 2, \ldots, r$, $U_i \in \{u_{i0}, u_{i1}, \ldots, u_{ik_i}\}$, whereas for the system it is assumed that $V \in \{v_0, v_1, \ldots, v_k\}$. For each component, the states are sorted in such a way that they are ranging from the worst (u_{i0}) to the best (u_{ik_i}) performance states. Same argument is valid for the system, where $k + 1$ different levels of performance are being considered ranging between complete failure (v_0) and perfect functioning (v_k). So $k_i + 1$ is the maximum number of states allowed for component i, whereas $k + 1$ is the maximum number of states that describe the behaviour of the system. Now, the structure function of the system is a mapping

$$\varphi : \{u_{10}, u_{12}, \ldots, u_{1k_1}\} \times \cdots \times \{u_{r0}, u_{r2}, \ldots, u_{rk_r}\} \mapsto \{v_0, v_2, \ldots, v_k\},$$

where $\times$ denotes the Cartesian product. Although, in the literature, this class of models is usually referred to as multi-state models, we prefer the present terminology (discrete structures) in order to give more emphasis to the distinction with the following class of models.

Continuum Structures In this case, we allow variables $U_1, U_2, \ldots, U_r$, and V to take any value in the interval $[0, 1]$. As mentioned before, there is no loss of generality if we assume 0 as the worst state for the system as well as for any component (*complete failure*), and 1 as the best state (*perfect functioning*). The state values between 0 and 1 are ordered according to performance, where higher number implies better performance.

If φ is the structure function, then again $V = \varphi(U_1, U_2, \ldots, U_r)$, and it is also assumed that $V \in [0, 1]$. So, the structure function of a continuous system is a mapping defined from the unit hypercube $\mathbf{S} = [0, 1]^r$ into the unit interval $[0, 1]$.

5.2 Monotone Regression Analysis of the Structure Function

In reliability theory, the state of a system, V, is expressed as a certain function of the state of its components, $\mathbf{U} = (U_1, U_2, \ldots, U_r)$, i.e. the structure function, $V = \varphi(\mathbf{U})$. This relationship is usually considered as deterministic. Nevertheless, and more and more because of the increase in the complexity of reliability systems, it would be more appropriate to assume that the behavior of all the components in the system may not be under control and specified by the vector $\mathbf{U}$. Therefore, it would be convenient to take into account some random perturbation in the specification of the structure function. Hence, for given observed data, $\{(\mathbf{U}_j; V_j) \in S \times [0, 1]; j = 1, 2, \ldots, n\}$, we can regard the data as being generated from the model $V = \varphi(\mathbf{U}) + \varepsilon$, that is, the system state is some function of $\mathbf{U}$ plus a random component error or noise, ε, called *residual*. The problem then is to fit an r-dimensional surface to the observed data that estimates $\varphi(\mathbf{u}) = E[V \mid \mathbf{U} = \mathbf{u}]$ for each given state $\mathbf{u}$. The error is often assumed to have $E[\varepsilon] = 0$ and variance σ^2, and $\mathbf{U}$ and ε are also assumed to be independent random variables.

We propose the use of nonparametric multivariate regression techniques for estimating the structure function, $\varphi(\mathbf{u})$. The estimation problem is formulated as follows:

Given a set of n observed data $\{(\mathbf{U}_j; V_j) \in S \times [0, 1]; j = 1, 2, \ldots, n\}$, the goal is to represent the structural relationship between the response variable V and the predictor vector $\mathbf{U}$ by a function, $\widetilde{\varphi}(\mathbf{U})$ that fits the data preserving the partial order. That is, given two vectors with $\mathbf{U}_{j_1} \prec \mathbf{U}_{j_2}$ then $\widetilde{\varphi}(\mathbf{U}_{j_1}) \leq \widetilde{\varphi}(\mathbf{U}_{j_2})$

Here, the symbol $\prec$ denotes the usual partial ordering defined in $\mathbf{S}$, i.e. $\mathbf{u}_1 \prec$ $\mathbf{u}_2 \Leftrightarrow u_{1i} \leq u_{2i}$ for all $i = 1, 2, ..., r$.

This formulation of the problem involves some important aspects to take into account. On the one hand, we are concerned with monotone regression in more than one explanatory (or predictor) variable. So, we are dealing with a minimization problem under monotonicity constraints induced by a partially ordered set of observations. On the other hand, to get more flexibility, we look for a function that fits our dataset without assuming any parametric model. The only assumption we make here is the smoothness of the target function (in the sense of differentiability).

In short, we want a multivariable function that is monotone in each argument and that models properly the relationship between V and $\mathbf{U}$. This problem can be broken down into two steps:

- First, we obtain a nonparametric multivariate regression estimator, $\widehat{\varphi}(\mathbf{u})$, for the expected value of V.
- And finally we find the monotone function, $\widetilde{\varphi}(\mathbf{u})$, closest to $\widehat{\varphi}(\mathbf{u})$.

5.2.1 Nonparametric Kernel Regression of the Structure Function

Classical parametric techniques in regression analysis assume that the regression function belongs to a certain parametric family of functions (such as linear functions, quadratic functions, etc.) and look for the best one inside the family. In many real situations, it is difficult, even impossible, to find such parametric function that represents the global relationship between V and $\mathbf{U}$. There are several methods that try to overcome these drawbacks of the parametric regression methods. Here, we are interested in local modelling methods, more specifically in the so-called *local (kernel) polynomial regression methods*. The local modelling approach relaxes the assumption of a global parametric functional form for the regression function. In fact, it is only assumed that the regression function is smooth (i.e. it has derivatives), which allows us to give a more flexible solution for the problem (see for example, [12], for a good description of these methods).

5.2.1.1 The Multivariate Local Linear Smoother

Suppose that $\varphi(\mathbf{u})$ has $k + 1$ derivatives at $\mathbf{u}$. Then, $\varphi(\mathbf{u})$ can be approximated in a neighborhood of $\mathbf{u}$ by a polynomial of order k, via a Taylor expansion. For simplicity, it is usually considered an approximation by the first-order polynomial

$$\varphi(\mathbf{u}^0) \approx \varphi(\mathbf{u}) + \nabla\varphi(\mathbf{u})(\mathbf{u}^0 - \mathbf{u}), \tag{5.1}$$

where $\nabla\varphi(\mathbf{u})$ is an r-dimensional vector that denotes the gradient of $\varphi(\mathbf{u})$. This polynomial can be fitted locally by the following weighted least squares regression problem

$$\min_{\alpha,\beta} \sum_{j=1}^{n}\{V_j - \alpha - (\mathbf{U}_j - \mathbf{u})\boldsymbol{\beta}\}^2 \mathbf{W}_h(\mathbf{u}; \mathbf{U}_j) \qquad (5.2)$$

with $(\alpha, \boldsymbol{\beta}^\top) = (\varphi(\mathbf{u}), \nabla\varphi(\mathbf{u}))$, where $^\top$ refers to the transpose operator. Here, the weights are defined locally for each observation being, thus depending on the considered estimation state, $\mathbf{u}$. They are defined by considering a multivariate kernel function that defines the shape of the weighting function into this local neighborhood. This kernel function is generally a symmetric probability density function, with compact support (although these conditions can be relaxed significantly). Hence, it is usually written that

$$\mathbf{W}_h(\mathbf{u}; \mathbf{U}_j) = h^{-1}\mathbf{W}\left(\frac{\mathbf{U}_j - \mathbf{u}}{h}\right), \qquad (5.3)$$

where $\mathbf{W} \geq 0$ and $\int \mathbf{W}(x_1, x_2, \ldots, x_r)\, dx_1 dx_2 \ldots dx_r = 1$. The parameter $h > 0$ is the bandwidth and controls the size of the local neighborhood around $\mathbf{u}$. For the sake of simplicity, we assume the same bandwidth for each component (a more general definition would involve a vector of bandwidths or even a matrix parameter).

By solving the minimization problem formulated in Expression (5.2), the estimation of the structure function at a given state vector $\mathbf{x}$ is obtained as

$$\widehat{\varphi}_h(\mathbf{u}) = \widehat{\alpha} = \mathbf{e}_{r+1}\left(\mathbf{U}_\mathbf{u}^\top \mathbf{D}_{\mathbf{u},h}\mathbf{U}_\mathbf{u}\right)^{-1}\mathbf{U}_\mathbf{u}^\top \mathbf{D}_{\mathbf{u},h}\mathbf{V}, \qquad (5.4)$$

where we denote $\mathbf{e}_{r+1} = (1, 0, \ldots, 0)_{1 \times (r+1)}$,

$$\mathbf{U}_\mathbf{u} = \begin{pmatrix} 1 & (\mathbf{U}_1 - \mathbf{u}) \\ \vdots & \vdots \\ 1 & (\mathbf{U}_n - \mathbf{u}) \end{pmatrix}_{n\times(r+1)}, \quad \mathbf{V} = \begin{pmatrix} V_1 \\ \vdots \\ V_n \end{pmatrix}_{n\times 1},$$

and $\mathbf{D}_{\mathbf{u},h}$ is an n-dimensional diagonal matrix whose entries are $\mathbf{W}_h(\mathbf{u};\mathbf{U}_j)$, for $j = 1, 2, \ldots, n$. The whole estimated surface is obtained with $\mathbf{u}$ varying in the estimation domain which, in this case, is the state space $\mathbf{S}$.

The estimator given in Eq. (5.4) is called Multivariate Local Linear Smoother (*MLLS*) and was proposed by Ruppert and Wand [28]. It may be expressed in the form

$$\widehat{\varphi}_h(\mathbf{u}) = \frac{1}{n}\sum_{j=1}^{n} \omega_j(\mathbf{u}; h)V_j, \qquad (5.5)$$

where the local weights, ω_j, depend only on the corresponding state vector $\mathbf{U}_j$, and, according to Expression (5.4), are given by

$$\omega_j(\mathbf{u}; h) = \mathbf{e}_{r+1}\left(\mathbf{U}_\mathbf{u}^\top \mathbf{D}_{\mathbf{u},h}\mathbf{U}_\mathbf{u}\right)^{-1}[1, \mathbf{U}_j - \mathbf{u}]^\top \mathbf{W}_h(\mathbf{u}; \mathbf{U}_j), \qquad (5.6)$$

with $[1, \mathbf{U}_j - \mathbf{u}]$ an $(r + 1)$-dimensional row-vector. The *MLLS* estimator has nice minimax efficiency properties and adapts automatically the bias at boundary points. The theoretical advantages of the *MLLS* jointly with its intuitive definition and its good performance in real applications lead the data analysts to choose this method rather than other nonparametric methods.

5.2.1.2 The Multivariate Nadaraya–Watson Smoother

An alternative solution for the regression problem is given by fitting the data (locally) by a polynomial of order 0 i.e. a constant. Hence, in the minimization problem stated in (5.1), it is established that $\beta = 0$. With such a simple approximation, the so-called Multivariate Nadaraya–Watson Smoother (*MNWS*) is defined. This estimator may also be expressed in the form (5.5), assigning weight

$$\omega_j(\mathbf{u}; h) = \frac{n\mathbf{W}_h(\mathbf{u}; U_j)}{\sum_{k=1}^{n} \mathbf{W}_h(\mathbf{u}; U_k)} \tag{5.7}$$

to the observation $\mathbf{U}_j, j = 1, 2, \ldots, n$. In its most popular form, this estimator is represented with

$$\omega_j(\mathbf{u}; h) = \frac{n\mathbf{W}\left(\frac{U_j - \mathbf{u}}{h}\right)}{\sum_{k=1}^{n} \mathbf{W}\left(\frac{U_k - \mathbf{u}}{h}\right)}, \tag{5.8}$$

since, as mentioned earlier, the kernel function, $\mathbf{W}$, is usually assumed to be a symmetric density function with compact support. The *MNWS* is a commonly used estimator because of its simplicity, which is more appreciated in the multi-dimensional framework that we are assuming here. However, an adequate performance of this latest estimator should come by making some good boundary correction. We deal with this problem by using a kernel correction in the following section.

5.2.2 The Problem of the Boundary Effect

Since components in the system are supposed to operate independently, the simpler is to use a multiplicative kernel function, although kernels without this property may be used. That is

$$\mathbf{W}_h(\mathbf{u}; U_j) = W_{1,h}(u_1; U_{j,1}) W_{2,h}(u_2; U_{j,2}) \ldots W_{r,h}(u_r; U_{j,r}), \tag{5.9}$$

for $j = 1, 2, \ldots, n$; $W_{i,h}$ being a univariate kernel function for each $i = 1, 2, \ldots, r$. For simplicity, we assume that $W_{i,h} = W_h$, for all $i = 1, 2, \ldots, r$. Usually, $W_h(\cdot) = h^{-1} W(\cdot/h)$, with W a symmetric density function. However, it turns out that a particularly serious problem arises when such functions are considered, that is, a significant bias error arises near the edges of the data (where this symmetric kernel function is truncated by the boundary). In order to correct boundary errors, we consider an asymmetric kernel function. In particular, we concentrate on the Beta family of distributions see [10], and therefore let $\widetilde{W}$ be the density function of a Beta distribution, i.e.

$$\widetilde{W}(u;a;b) = \frac{\Gamma(a+b)}{\Gamma(a)\Gamma(b)}u^{a-1}(1-u)^{b-1}, \qquad (5.10)$$

for $0 \leq u \leq 1$, with $a, b > 1$ the shape parameters. This specification of the Beta distribution ensures that the density function is unimodal. The mean and the variance of the Beta density, given a and b, are obtained, respectively, as

$$\mu = \frac{a}{a+b} \text{ and } \sigma^2 = \frac{ab}{(a+b)^2(a+b+1)}.$$

The mean value controls the center of the distribution, while the variance determines the scale, so that different values for σ^2 will control the smoothness of the kernel estimation. For our purposes, we consider a Beta kernel with parameters

$$a = \frac{u_0}{h} + 1 \text{ and } b = \frac{1-u_0}{h} + 1,$$

for $u_0 \in [0, 1]$ and the bandwidth parameter $h > 0$. Here, u_0 will be the mode of the distribution. Figure 5.1 illustrates several shapes of the Beta distribution for different choices of u_0 and h, which provides different combinations of parameters a and b, respectively, and therefore different shapes of the corresponding density curve.

Let $\widetilde{W}(\cdot; u_0/h + 1; (1 - u_0)/h + 1)$ denote the Beta kernel with mode u_0 and bandwidth h. Easy computations lead to the following expression for the variance

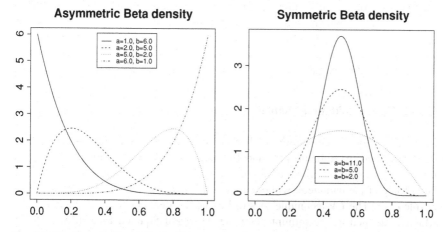

Fig. 5.1 The Beta family of distributions

$$\sigma^2 = \frac{h(u_0 + h)(1 - u_0 + h)}{(2h + 1)^2(3h + 1)} = \frac{h^3 + h^2 + hu_0(1 - u_0)}{12h^3 + 16h^2 + 7h + 1}$$

$$\approx \frac{hu(1 - u_0) + o(h^2)}{1 + o(h)} \approx hu_0(1 - u_0) + o(h^2)$$

which means that both u_0 and h determine the amount of smoothness involved by the Beta-kernel estimator.

Finally, we consider in Expression (5.9) the following Beta-kernel modified at the boundaries [10]

$$W_h(u; u_0) = \begin{cases} \widetilde{W}(u; \lambda(u_0, h); (1 - u_0)/h + 1) & \text{if } u_0 \in [0, 2h) \\ \widetilde{W}(u; u_0/h + 1; (1 - u_0)/h + 1) & \text{if } u_0 \in [2h, 1 - 2h], \\ \widetilde{W}(t; u_0/h + 1; \lambda(1 - u_0, h)) & \text{if } u_0 \in (1 - 2h, 1] \end{cases} \quad (5.11)$$

for each u_0 and given bandwidth h, where

$$\lambda(u_0, h) = 2h^2 + 2.5 - \sqrt{(2h^2 + 1.5) - u_0^2} - u_0/h.$$

For each sample value, U_j with $j = 1, 2, ..., n$, factors in Expression (5.9) are obtained evaluating the function (5.11) in each element of vector $\mathbf{u}$ and the one corresponding in the vector $\mathbf{U}_j$.

5.2.3 Data-Driven Bandwidth Selection Methods

The bandwidth parameter h determines the complexity of the local model. A nearly zero bandwidth results in an interpolation of the data i.e. the higher complexity. On the other hand, a bandwidth near to infinity corresponds to fitting globally a polynomial to the data. In this sense, the choice of the best bandwidth becomes a problem of compensating the bias and the variance of the resulting estimator. Apart from any subjective bandwidth choices, the usual way is to define the theoretical optimal bandwidth as the minimizer of some error criterion and try to estimate it from the data. Such an automatic method provides the so-called data-driven bandwidth estimators.

5.2.3.1 Cross-Validation Method

Probably the most used data-driven bandwidth is the least squares cross-validation (*CV*) proposed by Rudemo [27] and Bowman [6]. The popularity of *CV* comes from a simple and intuitive definition that makes the practitioners to use it for a fast choice of the parameter without assuming any difficult conditions as the plug-in rule does. The *CV* technique tries to imitate the theoretical optimal bandwidth, which is defined as the minimizer of the Averaged Squared Error (*ASE*) of the estimator $\widehat{\varphi}_h(\mathbf{u})$, given by

$$ASE(h) = \frac{1}{n} \sum_{j=1}^{n} \{ \varphi(\mathbf{U}_j) - \widehat{\varphi}_h(\mathbf{U}_j) \}^2. \tag{5.12}$$

To do so, we consider the *CV* function

$$CV(h) = \frac{1}{n} \sum_{i=1}^{n} \{ V_j - \widehat{\varphi}_{h,-j}(\mathbf{U}_j) \}^2 \tag{5.13}$$

in order to approximate the *ASE* given in (5.12). Then, we define the cross-validation bandwidth as the value of h that minimizes $CV(h)$. This bandwidth is called h_{CV} in the following. Here, $\widehat{\varphi}_{h,-j}(\cdot)$ denotes the nonparametric regression estimator that is obtained by leaving out the ith pair of the $(\mathbf{U}_j, V_j)$ observations.

5.2.3.2 Bootstrap Method

Other popular methods for selecting bandwidths are based on the Bootstrap method. The so-called Wild Bootstrap has recently been used by González-Manteiga et al. [14] in the context of multivariate local linear regression to select the bandwidth parameter. In that paper, the bootstrap bandwidth selector showed a good practical performance beating other competitors such as the CV bandwidth. Here, we also consider a bootstrap bandwidth selector to estimate the optimal bandwidth, in the sense of minimizing the Mean Squared Error (*MSE*) of the nonparametric regression estimator,

$$MSE(\mathbf{u}, h) = E[\varphi(\mathbf{u}) - \widehat{\varphi}_h(\mathbf{u})|\mathbf{U} = \mathbf{u}]^2. \tag{5.14}$$

The bootstrap bandwidth choice comes from an estimation of *MSE* based on the following resampling scheme.

Algorithm 5.1 (*Bootstrap Method*)
Step 1. Choose a pilot bandwidth h_0 and estimate the structure function via $\widehat{\varphi}_{h_0}(\mathbf{u})$;
Step 2. Estimate the residuals by $\widehat{\varepsilon}_j = V_j - \widehat{\varphi}_{h_0}(\mathbf{U}_j)$;
Step 3. Draw a bootstrap residual sample $\{ \widehat{\varepsilon}_1^*, \widehat{\varepsilon}_2^*, \ldots, \widehat{\varepsilon}_n^* \}$, under conditions

$$E_*\left[\widehat{\varepsilon}_j^* \right] = 0; \quad E_*\left[\widehat{\varepsilon}_j^{*2} \right] = \widehat{\varepsilon}_j^2; \quad E_*\left[\widehat{\varepsilon}_j^{*3} \right] = \widehat{\varepsilon}_j^3;$$

Step 4. Construct the bootstrap sample $\{ (\mathbf{U}_1, V_1^*), (\mathbf{U}_2, V_2^*), \ldots, (\mathbf{U}_n, V_n^*) \}$, by means of

$$V_j^* = \widehat{\varphi}_{h_0}(\mathbf{U}_j) + \varepsilon_j^*;$$

Step 5. Calculate the bootstrap nonparametric estimator $\widehat{\varphi}_h^*(\mathbf{u})$ similar to $\widehat{\varphi}_h(\mathbf{u})$, but based on the bootstrap sample generated at the previous step. Finally, define the bootstrap approximation of the *MSE* given in (5.14) as follows,

$$MSE^*(\mathbf{u}, h) = E_* \left[\left(\widehat{\varphi}_h^*(\mathbf{u}) - \widehat{\varphi}_{h_0}(\mathbf{u}) \right)^2 \right].$$

The expectation in Step 5 is computed over the resampling distribution defined at Step 3. The MSE^* can be exactly calculated. In fact, it can be written

$$MSE^*(\mathbf{u}, h) = \left(B_{h,h_0}^*(\mathbf{u}) \right)^2 + Var_{h,h_0}^*(\mathbf{u}), \qquad (5.15)$$

with bootstrap bias and bootstrap variance of (5.15) given, respectively, by

$$B_{h,h_0}^*(\mathbf{u}) = \sum_{j=1}^{n} \omega_j(\mathbf{u}, h) \widehat{\varphi}(\mathbf{U}_j, h_0) - \widehat{\varphi}(\mathbf{u}, h_0)$$

and

$$Var_{h,h_0}^*(\mathbf{u}) = \sum_{j=1}^{n} \omega_j(\mathbf{u}, h)^2 \widehat{\varepsilon}_j^2,$$

see González-Manteiga et al. [14] for further details.

Finally, the bootstrap bandwidth, which we denote by h_{boot}, is defined by minimizing Expression (5.15) over h, for each state vector $\mathbf{u}$. Note that this method provides a local bandwidth selector, i.e. a bandwidth which is a function of $\mathbf{u}$ instead of a constant value as the CV method does. The local nature of the bootstrap bandwidth provides a notable improvement on the fitting that is more appreciable for difficult regression estimation problems (spurious and irregular surfaces).

5.2.4 Estimation Under Monotonicity Constraints: Isotonization Procedure

The nonparametric regression estimators proposed in Sect. 5.2.1 are not necessarily monotone. Therefore, once the smooth estimator of the structure function has been obtained, we are going to proceed to isotonize this estimated function in each argument.

In other words, starting from a dataset $\{(\mathbf{U}_j, V_j) \in \mathbf{S} \times [0, 1]; j = 1, 2, \ldots, n\}$, we have already built a smooth estimator of type $\widehat{\varphi}_h(\mathbf{u})$. It is desirable that this function satisfies that $\mathbf{u}_1 \prec \mathbf{u}_2 \Rightarrow \widehat{\varphi}_h(\mathbf{u}_1) \leq \widehat{\varphi}_h(\mathbf{u}_2)$, where, $\prec$ means the partial order defined earlier. This property of monotonicity is not assured for the function estimated by means of the smoothing procedure we have just used, so we now try to find the monotone function, $\widetilde{\varphi}_h(\mathbf{u})$, closest to $\widehat{\varphi}_h(\mathbf{u})$.

When it is the case of a completely ordered set (e.g. $r = 1$), the Pool Adjacent Violators Algorithm ($PAVA$) developed by Ayer et al. [2] can provide a solution. We consider a generalized version of this algorithm that applies to partially ordered datasets (i.e. for the general case of $r > 1$).

The problem of simple isotonization can be formulated as follows.

Let $\{(x_j, y_j); j = 1, 2, \ldots, n\}$ be a two-dimensional dataset such that the x_j's are given in ascendant order. Find $\{f(x_j), j = 1, 2, \ldots, n\}$ to minimize $n^{-1}\sum_{j=1}^{n}(y_j - f(x_j))^2$ subject to $f(x_1) \leq f(x_2) \leq \cdots \leq f(x_n)$.

A solution can be obtained from the *PAVA*, which can be summarized in the two following steps.

Algorithm 5.2 (*PAVA*)

Step 1. Start with y_1. Move to the right and stop if the pair (y_j, y_{j+1}) violates the monotonicity constraint, that is, if $y_j > y_{j+1}$. Pool y_j and the adjacent y_{j+1}, by replacing them both by their average,

$$y_j^{\bullet} = y_{j+1}^{\bullet} = \frac{y_j + y_{j+1}}{2}.$$

Step 2. Next, check that $y_{j-1} < y_j^{\bullet}$. If not, pool (y_{j-1}, y_j, y_{j+1}) into their average. Move to the left until the monotonicity condition is fulfilled. Then proceed to the right. The final solution is $f(y_1) \leq f(y_2) \leq \cdots \leq f(y_n)$.

Next, we propose a solution for the multi-dimensional case, that is, we take advantage of the *PAVA* to find a solution for the function $\widetilde{\varphi}_h(\mathbf{u})$ satisfying the conditions above.

Firstly, define a grid of d points on the unit interval $[0, 1]$, that is $G = \{l/d; l = 1, 2, \ldots, d\}$, and consider $G^r = G \times G \times \cdot^r \cdot \times G$. Calculate the values of the estimator $\widehat{\varphi}_h(\mathbf{u})$ at each $\mathbf{u} \in G^r$. Given $r \geq 1$, the isotonization procedure for the r-dimensional case is solved based on the solution for the $(r - 1)$-dimensional case. In other words, fix a component of the vector $\mathbf{u}$, for instance u_1. Each value of u_1 generates a problem of isotonization of order $r - 1$. So, if the problem is solved for a lower dimension, then we can continue on. At the lowest stage, $r = 1$, we make use of the function *pava*, which is contained in the library *Iso* [30] of the environment R and implements *PAVA*.

5.2.5 Examples

In this section, we illustrate the proposed method in some practical situations and evaluate its performance through simulation experiments. Thereby, we will consider two examples of continuous-state systems. For both examples, we obtain an isotonized version of the multivariate local polynomial estimator proposed in previous sections.

5.2.5.1 Series System

Let us consider a series system with $r = 2$ independent components. Then

$$\varphi(u_1, u_2) = \min\{u_1, u_2\}$$

is the true structure function of the system, since the state of a continuous series system is *as bad as the one of its "worst" component.*

The simulation study is carried out under the following settings. We generate a sample of size $n = 100$, with $\mathbf{S} = [0, 1]^2$. The state of each component is generated randomly from the uniform distribution, but other possibilities are also acceptable. We consider the presence of a stochastic noise ε (residual) in the behavior of the system, in such a way that the resulting state of the system, V, is determined by the following relationship,

$$V = \min\{U_1, U_2\} + \varepsilon. \tag{5.16}$$

We assume for the residual ε a normal distribution with mean 0 and standard deviation 0.1. In order to eliminate the sample effect, we make 1,000 repetitions of the sampling model. For each sample we apply the estimation procedure developed in the former subsections. We consider both *MLLS* and *MNWS* estimators with the boundary-corrected kernel presented in Sects. 5.2.1 and 5.2.2. The bandwidth parameter, h, is estimated by the *CV* criterion (h_{CV}) and the local bootstrap method (h_{boot}) described in Sect. 5.2.3.

We present results concerning three different estimators. The first one, denoted *MNWS-Boot*, is obtained as a Nadaraya–Watson-type estimator as given in (5.8) with a bootstrap procedure conducted for the bandwidth estimation. The second one, denoted *MNWS-CV*, also refers to a Nadaraya–Watson-type estimator but now the *CV* method is considered for approximating the bandwidth. The third estimator is named *MLLS-CV* and is obtained as explained in Sect. 5.2.1.1. In this last case, the *CV* method is considered to derive bandwidth estimation. The combination *MLLS* and *bootstrap bandwidth* has also been considered. It is not included here because the results obtained do not improve significantly the ones reported by the third type of estimator described earlier, and, on the other hand, it implies an important increase in the computational burden.

The whole estimated surface is outlined by calculating the estimator (*MNWS-Boot, MNWS-CV* and *MLLS-CV*) on a grid of 100 equally spaced points in $\mathbf{S} = [0, 1]^2$ (denoted by $\mathbf{u}_l$, $l = 1, \ldots, 100$). To asses the practical performance of an estimator, we proceed as follows. We consider a replication of the model and construct the corresponding estimator from such a replication. Let $\tilde{\varphi}_h^m$ denote the nonparametric estimator of the structure function obtained from the m-th replication, for $m = 1, \ldots, 1,000$. Then, we consider the Integrated Squared Error (*ISE*) criterion defined by

$$ISE(m) = \frac{1}{100} \sum_{l=1}^{100} \left(\varphi(\mathbf{u}_l) - \tilde{\varphi}_h^m(\mathbf{u}_l)\right)^2, \tag{5.17}$$

see Eq. (5.12) to note the difference with the *ASE* criterion. We have carried out some statistical and graphical analyses concerning the estimation results obtained by means of the considered nonparametric estimators *MNWS-Boot, MNWS-CV* and *MLLS-CV*. Figure 5.2 displays the box-and-whisker diagrams (left panel) as well as the estimated density traces (right panel) of the obtained *ISE* values over the whole

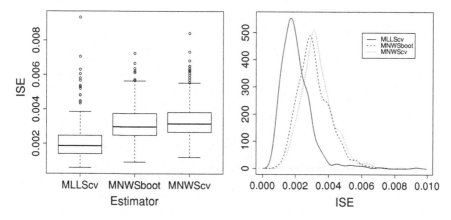

Fig. 5.2 Box-and-whisker plots (*left panel*) and estimated density curves (*right panel*) of the ISE values for the series system

Table 5.1 Summary statistics for the ISE's

	MLLScv	MNWSboot	MNWScv
Min	6.049362e-004	9.324737e-004	1.217009e-003
1st Qu.	1.411529e-003	2.483761e-003	2.672268e-003
Mean	2.033477e-003	3.115622e-003	3.324471e-003
Median	1.862617e-003	2.969054e-003	3.156129e-003
3rd Qu.	2.462066e-003	3.753746e-003	3.820079e-003
Max	9.336906e-003	7.238546e-003	8.412483e-003
Std. Dev.	9.468672e-004	9.663086e-004	9.897198e-004

set of replications. According to these graphs, the *MLLS*-type estimator seems to be the most appropriate. In fact, it beats clearly the *MNWS*-type estimator, even when the local bootstrap bandwidth is considered. These conclusions are corroborated by Table 5.1, which reports some descriptive statistics about the *ISE* values.

Furthermore, we have sorted the resultant estimated surfaces by its associated value of the *ISE* and, to summarize, we have only considered those ones with an *ISE* value above the 90th percentile. In Fig. 5.3, we present the corresponding estimated structure function for the two-component series system whose error has been measured as $ISE = 0.000605$. The solid circles plotted on the surface correspond to the estimated values of the function on a grid of 100 points selected in $S = [0, 1]^2$.

5.2.5.2   Two-Terminal Network System

Let us now consider a hybrid combination of series and parallel structure in the following sense. The structure function will be determined as the combination of a pure parallel system of order two plus a third component disposed in series with

Fig. 5.3 Estimated series
structure with continuous-
state space

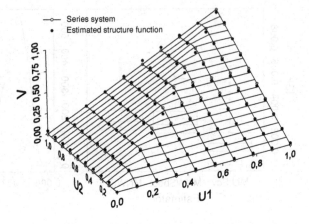

Fig. 5.4 Two-terminal
network system with three
components

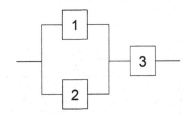

the rest. This results in a series–parallel system as the one displayed in Fig. 5.4. Such structures have been considered for instance in Sarhan [29] as an example of a radar system in an aircraft, and in Li et al. [19] as a two-terminal network.

Under the binary approach, the system consists of three independent components that have only two states (functioning and failure) and the structure function of the system is obtained as

$$\varphi(u_1, u_2, u_3) = \min\{\max\{u_1, u_2\}, u_3\}. \tag{5.18}$$

Let us now consider the continuous version of this structure function, that is, the states of the components as well as the state of the system are considered to be continuous random variables with support [0, 1]. Furthermore, let us assume that components 1 and 2 do not share any task. If we think of the system as a two-terminal network into a communication system context, we could understand the state of a component as its capacity of transmission. Hence, we assume that the sub-system composed by {1, 2} is functioning through that component with the highest flow capacity, being the other component in a passive state (see [20]). So, under our continuum scope, the state of this parallel system (with no sharing tasks) takes the value of the state of its best component, and then the structure function is calculated as in the expression (5.18).

For the empirical study, we generate $n = 100$ points in the component state space $S = [0, 1]^3$. The state of each component is generated randomly from the uniform distribution. The state of the system V is given by the following relation

$$V = \min\{\max\{U_1, U_2\}, U_3\} + \varepsilon, \tag{5.19}$$

where ε is again supposed to be normal distributed with mean equal to 0 and standard deviation, 0.1. We have carried out a simulation study similar to the one in the previous example. Here, we consider 100 replications of the model.

Now, we only consider the *MNWS* estimator with *CV* bandwidth and again we apply the error criterion *ISE* defined as earlier, for each generated sample. The 90th percentile of the sample obtained for the *ISE*s gives the nice value of $ISE = 0.006878027$. The results for the *MLLS* estimator will outperform this error level but with a more computational effort.

5.3 Estimation of Some Performance Measures

We consider now some performance measures of interest from a reliability study point of view. We show how to obtain approximate expressions for some relevant measures as the expected state and the probability of failure from the estimator of the structure function.

5.3.1 The Expected State of the System

To fix a context, we first define what we mean by the system expected state.

Definition 5.3 (*System Expected State (SES)*) Let us consider a reliability system with r independent components whose structure function is given by $\varphi(U_1, U_2, \ldots, U_r)$, U_i being the random state of the ith component. Let us denote by f_i, the *pdf* of the variable U_i, for $i = 1, 2, \ldots, r$. Then, the expected state of the system is defined as the expectation of its structure function, that is

$$\begin{aligned} SES = E[\varphi(\mathbf{U})] &= E[\varphi(U_1, U_2, \ldots, U_r)] \\ &= \int_S \varphi(u_1, u_2, \ldots, u_r) f_1(u_1) f_2(u_2) \cdots f_r(u_r) du_1 du_2 \cdots du_r, \end{aligned} \tag{5.20}$$

where we are denoting $\mathbf{S} = [0,1]^r$. Note that when only two states $(0, 1)$ are allowed in the system, that is for binary systems, the expression in (5.20) reduces to the probability that the system is occupying state 1. So, for non-repairable binary systems, it is equivalent to the reliability function, and when repairs are taken account in the system the function in (5.20) is equivalent to the availability function. The expression in (5.20) can be approximated by the following,

$$E[\varphi(\mathbf{U})] \approx E[\widetilde{\varphi}(\mathbf{U})], \tag{5.21}$$

where $\widetilde{\varphi}$ is an isotonic nonparametric estimator of φ.

The expression in the right hand of (5.21) is also unknown given that the density functions $f_1, f_2, \ldots, f_r$ involved are not known. One method to approximate this expression could be by using a cross-validation criterion (*CV*). Briefly explained, the method consists of holding out one observation (e.g. j), which is used as a *validation data*; then a model is fitted to the remaining data that are called the *training set* (e.g. $\widetilde{\varphi}_{h,-j}$), and the fitted model is used to predict for the validation data. This procedure is repeated for each single data. The average of the resulting predictions is taken as the estimator of the above expectation in (5.20), that is

$$\widehat{SES} = \frac{1}{n}\sum_{j=1}^{n} \widetilde{\varphi}_{h,-j}(\mathbf{U}_j) = \frac{1}{n}\sum_{j=1}^{n}\left[\frac{1}{n-1}\sum_{\substack{l=1 \\ l\neq j}}^{n} \omega_l(\mathbf{U}_j, h)V_l\right], \qquad (5.22)$$

see Härdle et al. [16].

Example Series System

Continuing with the example in Sect. 5.2.5.1, we have estimated the expected state of this system by the procedure that we have just described. As reasoned earlier, for the series system, the state of the system is given by the minimum of the states of its components, so, the state of the system $\varphi(\mathbf{U}) = \min\{U_1, U_2\}$. If U_1 and U_2 are both uniformly distributed in the unit interval, then, the reliability function of the random variable $\varphi(\mathbf{U})$ is given by

$$\bar{F}_{\varphi}(v) = Pr\{\varphi(\mathbf{U}) > v\} = \begin{cases} 1, & v < 0; \\ (1-v)^2, & 0 \leq v \leq 1; \\ 0, & v > 1. \end{cases} \qquad (5.23)$$

Therefore, the state expectation is obtained as $E[\varphi(\mathbf{U})] = 1/3$.

We have considered the results obtained by means of the estimator of the structure function denoted by *MLLS-CV* and obtained previously in Sect. 5.2.5.1. The expected value of the structure function has been approximated according to the *CV* criterion (5.22) from each of the 1,000 samples generated for the model. To summarize the results, we consider the average value of the state expectation estimates along the 1,000 replications, which gives a value of $\widehat{SES} = 0.3410309$. The scatterplot corresponding to the replications is given in Fig. 5.5.

Example Two-terminal Network

Based on the results of Sect. 5.2.5.2, we have obtained an estimate of the expected state for this system (again using the *MLLS-CV* estimator for the structure function). The true expected value for the state of the system, that is $E[\varphi(U_1, U_2, U_3)]$, is obtained by integrating between 0 and 1 the expression $\bar{F}_{\varphi}(v) = 1 - v - v^2 + v^3$, the reliability function of the random variable state of the system. Hence, we have that $E[\varphi(U_1, U_2, U_3)] = 0.4166667$. We have applied the former *CV* procedure (5.22) and have obtained a point estimation of the expected structure

Fig. 5.5 Scatterplot of the estimations of the expected state of the system for 1,000 replications of the system

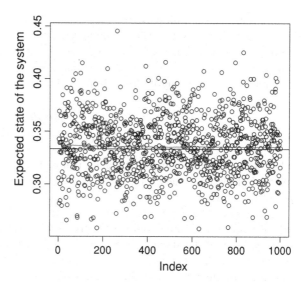

function given by $E[\widetilde{\varphi}(U_1, U_2, U_3)] \approx 0.4165224$. The number of replications carried out in this case was 100.

5.3.2 The Probability of Occurrence of Failures

One of the most interesting reliability indexes is given by the following measure:

$$P_\vartheta = Pr\{\varphi(\mathbf{U}) > \vartheta\}, \tag{5.24}$$

for any fixed $0 < \vartheta < 1$, which can be interpreted as the probability that the system gives a better performance than the one represented by state ϑ.

To estimate this probability, we can consider it as an expectation and use the arguments of the previous section. That is, $Pr\{\varphi(\mathbf{U}) > \vartheta\} = E(I[\varphi(\mathbf{U}) > \vartheta])$, with $I[\cdot]$ the indicator function. Then, if function φ is approximated by an isotonic nonparametric estimator of type described earlier, $\widetilde{\varphi}$, we can make $Pr\{\varphi(\mathbf{U}) > \vartheta\} \approx Pr\{\widetilde{\varphi}(\mathbf{U}) > \vartheta\}$ and reasoning as in Sect. 5.3.1,

$$E(I[\widetilde{\varphi}_h(\mathbf{U}) > \vartheta]) \approx \frac{1}{n}\sum_{j=1}^{n} I\left[\widetilde{\varphi}_{h,-j}(\mathbf{U}_j) > \vartheta\right] = \frac{1}{n}\{\sharp j : \widetilde{\varphi}_{h,-j}(\mathbf{U}_j) > \vartheta\}. \tag{5.25}$$

Then, P_ϑ can be estimated by

$$\widehat{P}_\vartheta = \frac{1}{n}\left\{\sharp j : \left(\frac{1}{n-1}\sum_{\substack{l=1 \\ j \neq j}}^{n} \omega_l(\mathbf{U}_j, h)V_j\right) > \vartheta\right\}. \tag{5.26}$$

If ϑ represents a limiting state of the system in the sense that if any lower state is achieved by the system some maintenance policy must be initiated, the probability of not exceeding such a limiting state ϑ acquire a key relevance. In fact, in structural reliability analysis, a fundamental task is to find a solution to the multifold integral representing the probability of failure (*POF*), given by

$$P_f = \int_{g(\mathbf{U})<0} f(\mathbf{u})d\mathbf{u}, \qquad (5.27)$$

where $\mathbf{U}$ is a vector of random variables that represent uncertain quantities such as material properties, loads, etc.; $g(\mathbf{U})$ is the performance function; and, $f(\mathbf{u})$ is the *pdf* of the vector $\mathbf{U}$.

The region where $g(\mathbf{U}) > (<)0$ is denominated the *safe (failure) region*. In practice, it is usually impossible to construct the joint density function $f(\mathbf{u})$ because of the scarcity of statistical data. Moreover, in case that statistical data is sufficient to form a joint distribution function, it may be quite difficult to perform the multidimensional integration over the generally irregular domain $\{g(\mathbf{U}) < 0\}$. For this task, it is usual to consider methods based on numerical analysis as well as Monte Carlo simulation [17, 31]. However, these simulation methods have revealed serious drawbacks that have motivated in the last two decades the development of the theory of a First-Order Reliability Method (*FORM*) and Second-Order Reliability Method (*SORM*), see for instance, Rackwitz [26] for a review.

The method proposed in this section provides a new perspective to face the problem. More specifically, the performance function g may be interpreted as a structure function φ, the vector $\mathbf{u}$ is similar to the vector state we consider, and the safe (failure) region could be the domain of $\mathbf{u}$ where $\varphi(\mathbf{u})$ takes values below the critical ϑ. Therefore, the *POF* could be expressed as $P_f = Pr_{\mathbf{U}}\{\varphi(\mathbf{U}) < \vartheta\}$. To

Fig. 5.6 Estimation of the expected probability of exceeding a given state value

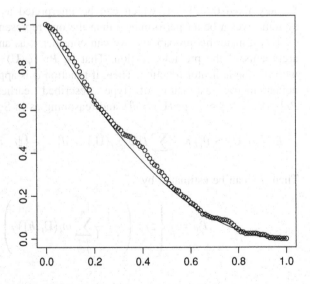

estimate this probability, we first obtain an estimate of the state function and then use a cross-validation method to approximate a value of the *POF* besides the expected state of the system.

Example Series System

We consider once again the series system introduced in Sect. 5.2.5.1. The empirical results are summarized in Fig. 5.6.

The points in the plot represent the estimation of the probability given in (5.24), i.e. $Pr\{\varphi(\mathbf{U}) > \vartheta_k\}$, for a grid of ϑ-values $\{\vartheta_k = k/100; \ k = 0, 1, ..., 100\}$. The solid line represents the theoretical distribution of the random state of the series system, that is the $Pr\{\varphi(\mathbf{U}) > \vartheta\}$, for $0 \leq \vartheta \leq 1$, which is explicitly given in Eq. (5.23).

Acknowledgments This chapter is an extension into book-length form of the article *Regression analysis of the structure function for reliability evaluation of continuous-state system*, originally published in *Reliability Engineering and System Safety* **95**(2), 134–142 (2010). The authors express their full acknowledgment of the original publication of the paper in the journal cited above, edited by Elsevier.

References

1. Aven T (1993) On performance measures for multistate monotone systems. Reliab Eng Syst Safety 41:259–266
2. Ayer M, Brunk HD, Ewing GM, Reid WT, Silverman E (1955) An empirical distribution function for sampling with incomplete information. Ann Math Stat 26:641–647
3. Barlow RE, Proschan F (1975) Statistical theory of reliability and life testing. Holt, Rinehart and Winston, New York
4. Baxter LA (1984) Continuum structures I. J Appl Prob 21:802–815
5. Baxter LA (1986) Continuum structures II. Math Proc Cambridge Philos Soc 99:331–338
6. Bowman A (1984) An alternative method of cross validation for the smoothing of density estimates. Biometrika 71:353–360
7. Brunelle RD, Kapur KC (1998) Continuous state system reliability: an interpolation approach. IEEE Trans Reliab 47(2):181–187
8. Brunelle RD, Kapur KC (1999) Review and classification or reliability measures for multistate and continuum models. IIE Trans 31:1171–1180
9. Burdakow O, Grimwall A, Hussian M (2004) A generalised PAV algorithm for monotonic regression in several variables. COMPSTAT'2004 Symposium
10. Chen SX (1999) Beta kernel estimators for density functions. Comput Stat Data Anal 31:131–145
11. El Neweihi E, Proschan F, Sethuraman J (1978) Multistate coherent systems. J Appl Prob 15(4):675–688
12. Fan J, Gijbels I (1996) Local polynomial modelling and its applications. Monographs on statistics and applied probability. Chapman & Hall, London
13. Gámiz ML, Martínez-Miranda MD (2010) Regression analysis of the structure function for reliability evaluation of continuous-state system. Reliab Eng Syst 95(2):134–142
14. González-Manteiga W, Martínez-Miranda MD, Pérez-González A (2004) The choice of smoothing parameter in nonparametric regression through wild bootstrap. Comput Stat Data Anal 47:487–515

15. Hall P, Hwang LS (2001) Nonparametric kernel regression subject to monotonicity constraints. Ann Stat 29(3):624–647
16. Härdle W, Müller M, Sperlich S, Werwatz A (2004) Nonparametric and semiparametric Models. Springer, Berlin
17. Kaymaz I, McMahon C (2005) A response surface method based on weighted regression for structural reliability analysis. Prob Eng Mech 20:11–17
18. Levitin G (2005) The universal generating function in reliability analysis and optimization. Springer, London
19. Li JA, Wu Y, Lai KK, Liu K (2005) Reliability estimation and prediction of multi state components and coherent systems. Reliab Eng Syst Safety 88:93–98
20. Lisnianski A (2001) Estimation of boundary points for continuum-state system reliability measures. Reliab Eng Syst Safety 74:81–88
21. Lisnianski A, Levitin G (2003) Multi-state system reliability. Assessment, optimization and applications. World Scientific, Singapore
22. Meng FC (2005) Comparing two reliability upper bounds for multistate systems. Reliab Eng Syst Safety 87:31–36
23. Mukarjee H, Stem S (1994) Feasible nonparametric estimation of multiargument monotone functions. J Am Stat Assoc 89:77–80
24. Natvig B (1982) Two suggestions of how to define a multistate coherent system. Appl Prob 14:434–455
25. Pourret O, Collet J, Bon JL (1999) Evaluation of the unavailability of a multistate component system using a binary model. Reliab Eng Syst Safety 64:13–17
26. Rackwitz R (2001) Reliability analysis a review and some perspectives. Struct Safety 23:365–395
27. Rudemo M (1982) Empirical choice of histograms and kernel density estimators. Scand J Stat 9:65–78
28. Ruppert D, Wand MP (1994) Multivariate locally weighted least squares regression. Ann Stat 22:1346–1370
29. Sarhan AM (2002) Reliability equivalence with a basic series/parallel system. Appl Math Comput 132:115–133
30. Turner R (2009) Iso: functions to perform isotonic regression http://CRAN.R-project.org/package=Iso
31. Wang L, Grandhi RV (1996) Safety index calculations using intervening variables for structural reliability analysis. Comput Struct 59(6):1139–1148

Chapter 6
Reliability of Semi-Markov Systems

6.1 Introduction

Semi-Markov processes constitute a generalization of Markov and renewal processes. Jump Markov processes, Markov chains, renewal processes—ordinary, alternating, delayed, and stopped—are particular cases of semi-Markov processes. As Feller [17] pointed out, the basic theory of semi-Markov processes was introduced by Pyke [54, 55]. Further significant results are obtained by Pyke and Schaufele [56, 57], Çinlar [11, 12], Koroliuk [32, 33, 34, 35], and many others. Also, see [3, 13, 19, 21, 23–26, 28, 29, 40, 42, 63]. Currently, semi-Markov processes have achieved significant importance in probabilistic and statistical modeling of real-world problems. Our main references for the present chapter are [6, 43, 50].

Reliability and related measurements, as availability, maintainability, safety, failure rate, mean times, etc., known under the term *dependability*, are very important in design, development, and life of real technical systems. From a mathematical point of view, the problems related to reliability are mostly concerned by the hitting time of a so-called failed or down subset of states of the system [6, 27, 31, 33, 43, 48].

Hitting times are important problems in theory and applications of stochastic processes. We encounter their distributions as reliability function in technical systems, or survival function in biology and medicine, or ruin probability in insurance and finance, etc. In this chapter, we are mostly concerned by reliability, but for the other above-mentioned problems, one can use the same models and formulae presented here. For general results on hitting times in semi-Markov setting, see e.g., [33, 61].

Reliability is concerned with the time to failure of a system which is the time to hitting the down set of states in the multi-state systems reliability theory [5, 8, 36, 46, 63]. In the case of repairable systems, this is the time to the first failure. Here, we are concerned by both repairable and nonrepairable systems. Apart from reliability,

M. L. Gámiz et al., *Applied Nonparametric Statistics in Reliability*,
Springer Series in Reliability Engineering, DOI: 10.1007/978-0-85729-118-9_6,
© Springer-Verlag London Limited 2011

we are concerned also by several related measurements as availability, maintainability, mean times to failure.

For example, when the temporal evolution of a system is described by a stochastic process, say Z, with state space E, it is necessary in reliability to define a subset of E, say D, including all failed states of the system. Then, the lifetime of such a system is defined by

$$T := \inf\{t \geq 0 : Z_t \in D\}.$$

The reliability is then defined as usual by $R(t) := P(T > t)$.

Of course, such a definition means that the system is binary, i.e., each state has to be an up or a down state. No degradation is considered. So, we consider the standard reliability model with catalectic failure modes (i.e., sudden and complete). For example, for a two-component parallel system, which is in up state when at least one of the two components is in up state, one considers four states: $12, \overline{1}2, 1\overline{2}, \overline{1}\overline{2}$, where i means up state and $\overline{i}$ means down state of component $i = 1, 2$. The degradation states $\overline{1}2, 1\overline{2}$ together with the perfect state 12 are considered to be up states. Nevertheless, by different definitions of the set D, for the same system, one can cover several cases of degradation by the same formulae presented here.

Stochastic processes are used from the very beginning of the reliability theory and practice. Markov processes, especially birth and death processes, Poisson processes, but also semi-Markov processes, especially the renewal processes, and some more general ones are used nowadays [6, 7, 8, 16, 31, 36, 43, 53].

The use of such stochastic processes in reliability studies is usually limited to finite state space, see e.g., [6, 10, 14, 30, 46, 50–53]. But in many real cases, on the one hand, the finite case, even the countable case, is not enough in order to describe and model the reliability of a real system. For example, in many cases, the state space can be $\{0,1\}^{N}$, in communication systems, or $\mathbf{R}_+ := [0,\infty)$ in fatigue crack growth modeling, e.g., [9, 15]. On the other hand, the Markov case is a restricted one since the lifetime of many industrial components as the mechanical ones are not exponentially distributed but mostly they are Weibull, log-normal, etc. In maintenance, the used distributions for durations are also not exponential distributed, but mostly log-normal and also fixed duration time are considered. Nowadays, an increasing interest for semi-Markov modeling in reliability studies of real-life systems is observed.

This is why we present a systematic modeling of reliability measurements also in the framework of semi-Markov processes. Particular cases of this modeling are the discrete state space, the discrete time, and the Markov and renewal processes (see [6, 43, 46]). While the discrete time case can be obtained from the continuous one, by considering countable measure for discrete time points, we considered that it is important to give separately this case since an increasing interest is observed in practice for the discrete case (see e.g., [6, 10, 46]).

This chapter is organized as follows. Section 6.2 presents some definitions and notation of semi-Markov processes in countable state space which are needed in the sequel. Further, in this section, some estimation problems are also given in the finite case for the semi-Markov kernel and the transition function of the semi-Markov process. Section 6.3 presents the continuous time reliability modeling. The reliability, availability, failure rate, ROCOF, and mean times are then defined into the semi-Markov setting, and their closed-form solutions are obtained via Markov renewal equation. Section 6.4 presents reliability in discrete time of semi-Markov chain systems. Section 6.5 is devoted to the semi-Markov processes with general state space toward reliability modeling. Finally, Sect. 6.6 presents stationary probability estimation and application to availability.

6.2 Markov Renewal and Semi-Markov Processes

6.2.1 Definitions

Consider an infinite countable set, say E, and an E-valued pure jump stochastic process $Z = (Z_t)_{t \in \mathbf{R}_+}$ continuous to the right and having left limits in any time point. Let $0 = S_0 \leq S_1 \leq \cdots \leq S_n \leq S_{n+1} \leq \cdots$ be the jump times of Z and $J_0, J_1, J_2, \ldots$ the successive visited states of Z. Note that S_0 may also take positive values. Let $\mathbf{N}$ be the set of nonnegative integers and $\mathbf{R}_+ := [0, \infty)$ the nonnegative real numbers.

Definition 6.1 (*Markov Renewal Process*) The stochastic process $(J_n, S_n)_{n \in \mathbf{N}}$ is said to be a Markov renewal process (MRP), with state space E, if it satisfies a.s., the following equality

$$P(J_{n+1} = j, S_{n+1} - S_n \leq t \mid J_0, \ldots, J_n; S_1, \ldots, S_n) = P(J_{n+1} = j, S_{n+1} - S_n \leq t \mid J_n)$$

for all $j \in E$, all $t \in \mathbf{R}_+$ and all $n \in \mathbf{N}$. In this case, Z is called a semi-Markov process (SMP).

We assume that the above probability is independent of n and S_n, and in this case, the MRP is called *time homogeneous*. The MRP $(J_n, S_n)_{n \in \mathbf{N}}$ is determined by the transition kernel $Q_{ij}(t) := P(J_{n+1} = j, S_{n+1} - S_n \leq t \mid J_n = i)$, called the *semi-Markov kernel*, and the *initial distribution* α, with $\alpha(i) = P(J_0 = i)$, $i \in E$. The process (J_n) is a Markov chain with state space E and transition probabilities $P(i,j) := Q_{ij}(\infty) := \lim_{t \to \infty} Q_{ij}(t)$, called the embedded Markov chain (EMC) of Z. It is worth noticing that here $Q_{ii}(t) \equiv 0$, for all $i \in E$, but in general we can consider semi-Markov kernels by dropping this hypothesis.

The semi-Markov process Z is connected to (J_n, S_n) by

$$Z_t = J_n, \quad \text{if} \quad S_n \leq t < S_{n+1}, \quad t \geq 0 \quad \text{and} \quad J_n = Z_{S_n}, \quad n \geq 0.$$

A Markov process with state space $E \subseteq \mathbf{N}$ and generating matrix $A = (a_{ij})_{i,j \in E}$ is a special semi-Markov process with semi-Markov kernel (Fig. 6.1)

$$Q_{ij}(t) = \frac{a_{ij}}{a_i}(1 - e^{-a_i t}), \quad i \neq j, \quad a_i \neq 0,$$

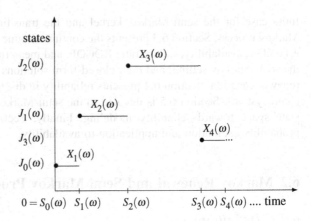

Fig. 6.1 A typical semi-Markov sample

where $a_i := -a_{ii}$, $i \in E$, and $Q_{ij}(t) = 0$, if $i = j$ or if $a_i = 0$.

Also, define $X_n := S_n - S_{n-1}$, $n \geq 1$, the interjump times, and the process $(N(t))_{t \in \mathbb{R}_+}$, which counts the number of jumps of Z in the time interval $(0, t]$, by

$$N(t) := \sup\{n \geq 0 : S_n \leq t\}.$$

Also, define $N_i(t)$ to be the number of visits of Z to state $i \in E$ in the time interval $(0, t]$. To be specific,

$$N_i(t) := \sum_{n=1}^{N(t)} \mathbf{1}_{\{J_n=i\}} = \sum_{n=1}^{\infty} \mathbf{1}_{\{J_n=i, S_n \leq t\}}, \quad \text{and also} \quad N_i^*(t) = \mathbf{1}_{\{J_0=i, S_0 \leq t\}} + N_i(t).$$

If we consider the (eventually delayed) renewal process $(S_n^i)_{n \geq 0}$ of successive times of visits to state i, then $N_i(t)$ is the counting process of renewals. Denote by μ_{ii} the mean recurrence time of the state i of Z. This is the mean interarrival times of the eventual detailed renewal process (S_n^i), $n \geq 0$, i.e., $\mu_{ii} = \mathrm{E}[S_2^i - S_1^i]$. In case where $Z_0 = i$, we have $S_0^i = 0$ and the renewal process (S_n^i) is an ordinary one; otherwise, if $Z_0 \neq i$, it is a delayed renewal process.

Let us denote by $Q(t) = (Q_{ij}(t), i, j \in E)$, $t \geq 0$, the semi-Markov kernel of Z. Then, we can write:

$$Q_{ij}(t) := P(J_{n+1} = j, X_{n+1} \leq t \mid J_n = i) = P(i,j)F_{ij}(t), \quad t \geq 0, i, j \in E, \quad (6.1)$$

where $P(i,j) := P(J_{n+1} = j \mid J_n = i)$ is the transition kernel of the EMC (J_n), and

$$F_{ij}(t) := P(X_{n+1} \leq t \mid J_n = i, J_{n+1} = j)$$

is the conditional distribution function of the sojourn time in the state i given that the next visited state is j, $(j \neq i)$. Let us also define the distribution function

$$H_i(t) := \sum_{j \in E} Q_{ij}(t)$$

and its mean value m_i, which is the mean sojourn time of Z in state i. In general, Q_{ij} is a sub-distribution, i.e., $Q_{ij}(\infty) \leq 1$; hence, H_i is a distribution function, $H_i(\infty) = 1$, and $Q_{ij}(0 -) = H_i(0 -) = 0$.

A special case of semi-Markov processes is the one where $F_{ij}(\cdot)$ does not depend on j, i.e., $F_{ij}(t) \equiv F_i(t) \equiv H_i(t)$ and

$$Q_{ij}(t) = P(i,j)H_i(t). \tag{6.2}$$

Any general semi-Markov process can be transformed into one of the form (6.2), (see e.g., [43]).

Let $\phi(i, t)$, $i \in E$, $t \geq 0$, be a real-valued measurable function and define the convolution of ϕ by Q as follows

$$Q * \phi(i,t) := \sum_{k \in E} \int_0^t Q_{ik}(ds)\phi(k, t - s). \tag{6.3}$$

Now, consider the n-fold convolution of Q by itself. For any $i, j \in E$,

$$Q_{ij}^{(n)}(t) = \begin{cases} \sum_{k \in E} \int_0^t Q_{ik}(ds)Q_{kj}^{(n-1)}(t - s) & n \geq 2 \\ Q_{ij}(t) & n = 1 \\ \delta_{ij}\mathbf{1}_{\{t \geq 0\}} & n = 0, \end{cases}$$

where δ_{ij} is the Kronocker symbol, $\delta_{ij} = 1$, if $i = j$ and zero otherwise. It is easy to prove (e.g., by induction) the following fundamental equality

$$Q_{ij}^{(n)}(t) = \mathrm{P}_i(J_n = j, S_n \leq t). \tag{6.4}$$

Here, as usual, $\mathrm{P}_i(\cdot)$ means $\mathrm{P}(\cdot \mid J_0 = i)$, and E_i is the corresponding expectation.

Let us define the Markov renewal function $\psi_{ij}(t)$, $i, j \in E$, $t \geq 0$, by

$$\psi_{ij}(t) := \mathrm{E}_i[N_j^*(t)] = \mathrm{E}_i \sum_{n=0}^{\infty} \mathbf{1}_{\{J_n=j,S_n \leq t\}}$$
$$= \sum_{n=0}^{\infty} \mathrm{P}_i(J_n = j, S_n \leq t) = \sum_{n=0}^{\infty} Q_{ij}^{(n)}(t). \tag{6.5}$$

Another important function is the semi-Markov transition function

$$P_{ij}(t) := \mathrm{P}(Z_t = j \mid Z_0 = i), \quad i, j \in E, t \geq 0,$$

which is the conditional marginal law of the process. We will study this function in the next section.

We will assume in the sequel that the semi-Markov process Z is regular, that is,

$$\mathrm{P}_i(N(t) < \infty) = 1,$$

for any $t \geq 0$ and any $i \in E$. And also that no distribution $F_i(t)$, $i \in E$ is a degenerate one.

For regular semi-Markov processes, we have $S_n < S_{n+1}$, for any $n \in \mathbf{N}$, and $S_n \to \infty$. The following theorem gives two criteria for regularity.

Theorem 6.1 *A semi-Markov process is regular if one of the following conditions is satisfied*:

(1) ([54]) *for every sequence* $(j_0, j_1, \ldots) \in E^\infty$ *and every* $C > 0$, *at least one of the series*

$$\sum_{k \geq 0} [1 - F_{j_k j_{k+1}}(C)], \qquad \sum_{k \geq 0} \int_0^C t F_{j_k j_{k+1}}(dt),$$

diverges;
(2) ([59]) *there exist constants, say* $\alpha > 0$ *and* $\beta > 0$, *such that* $H_i(\alpha) < 1 - \beta$, *for all* $i \in E$.

Example Alternating Renewal Process

Let us consider an alternating renewal process with lifetime and repair time distributions: F and G

- Up times: $X_1', X_2', \ldots$
- Down times: $X_1'', X_2'', \ldots$

Denote by S_n the starting (arrival) time of the $(n + 1)$-th cycle, that is

$$S_n = \sum_{i=1}^n (X_i' + X_i''), \quad n \geq 1.$$

The process

$$Z_t = \sum_{n \geq 0} \mathbf{1}_{\{S_n \leq t < S_n + X_{n+1}'\}}, \quad t \geq 0,$$

is a semi-Markov process, with states:

- 1 for functioning,
- 0 for failure and semi-Markov kernel:

$$Q(t) = \begin{pmatrix} 0 & F(t) \\ G(t) & 0 \end{pmatrix}$$

The embedded Markov chain (J_n) is a deterministic chain with transition matrix

$$P = \begin{pmatrix} 0 & 1 \\ 1 & 0 \end{pmatrix}$$

and then we have the corresponding MRP (J_n, S_n).

Let us now discuss the nature of different states of an MRP. An MRP is irreducible, if, and only if, its EMC (J_n) is irreducible. A state i is recurrent (transient) in the MRP, if, and only if, it is recurrent (transient) in the EMC. For an irreducible finite MRP, a state i is positive-recurrent in the MRP, if, and only if, it is recurrent in the EMC and if for all $j \in E$, $m_j < \infty$. If the EMC of an MRP is irreducible and recurrent, then all the states are positive-recurrent, if, and only if, $m := vm := \sum_i v_i m_i < \infty$ and null-recurrent, if, and only if, $m = \infty$ (where v is the stationary probability of EMC (J_n)). A state i is said to be periodic with period $a > 0$ if $G_{ii}(\cdot)$ (the distribution function of the random variable $S_2^i - S_1^i$) is discrete concentrated on $\{ka : k \in \mathbf{N}\}$. Such a distribution is said to be also periodic. In the opposite case, it is called aperiodic. Note that the term *period* has a completely different meaning from the corresponding one of the classic Markov chain theory.

6.2.2 Markov Renewal Theory

As the renewal equation in the case of the renewal process theory on the half-real line, the Markov renewal equation is a basic tool in the theory of semi-Markov processes.

Let us write the Markov renewal function (6.5) in matrix form

$$\psi(t) = (I - Q(t))^{(-1)} = \sum_{n=0}^{\infty} Q^{(n)}(t). \tag{6.6}$$

This can also be written as

$$\psi(t) = I(t) + Q * \psi(t), \tag{6.7}$$

where $I := I(t)$ (the identity matrix), if $t \geq 0$ and $I(t) = 0$, if $t < 0$. The upper index (-1) in the matrix $(I - Q(t))$ means its inverse in the convolution sense.

Equation (6.7) is a special case of what is called a *Markov Renewal Equation* (MRE). A general MRE is as follows

$$\Theta(t) = L(t) + Q * \Theta(t), \tag{6.8}$$

where $\Theta(t) = (\Theta_{ij}(t))_{i,j \in E}, L(t) = (L_{ij}(t))_{i,j \in E}$ are matrix-valued measurable functions, with $\Theta_{ij}(t) = L_{ij}(t) = 0$ for $t < 0$. The function $L(t)$ is a given matrix-valued function and $\Theta(t)$ is an unknown matrix-valued function. We may also consider a vector version of Eq. (6.8), i.e., consider corresponding columns of the matrices Θ and L.

Let us consider a function V defined on $E \times \mathbf{R}$ which satisfy the following two properties:

- $V(x,t) = 0$ for $t \leq 0$ and any $x \in E$;
- it is uniformly bounded on E on every bounded subset of $\mathbf{R}_+$. That is

$$\|V\|_{\infty,t} := \sup_{(x,s)\in E\times[0,t]} |V(x,s)| < +\infty,$$

for every $t > 0$. Denote by $\mathbf{B}_1$ the set of all such functions.

We say that a matrix function $\Theta(t) = \Theta_{ij}(t)$ belongs to $\mathbf{B}_1$, if for any fixed $j \in E$ the column vector function $\Theta_{.j}(\cdot)$ belongs to $\mathbf{B}_1$.

Equation (6.8) has a unique solution $\Theta = \psi * L(t)$ belonging to $\mathbf{B}_1$ when $L(t)$ belongs to $\mathbf{B}_1$.

The following assumptions are needed for results presented here.

Assumptions 6.1

A0: The semi-Markov process is regular.

A1: The stochastic kernel $P(i, j) = Q_{ij}(\infty)$ induces an irreducible ergodic Markov chain with the stationary distribution $v := (v_i, i \in E)$, and the mean sojourn times are uniformly bounded, that is:

$$m_i := \int_0^\infty \overline{H}_i(t)dt \leq C < +\infty,$$

$$m := \sum_{i\in E} v_i m_i > 0.$$

Theorem 6.2 (Markov Renewal Theorem [60])) *Under Assumptions A0–A1 and the following assumption:the functions $L_{ij}(t)$, $t \geq 0$, are direct Riemann integrable, i.e., they satisfy the following two conditions, for any $i, j \in E$:*

$$\sum_{n\geq 0} \sup_{n\leq t\leq n+1} |L_{ij}(t)| < \infty,$$

and

$$\lim_{\Delta\downarrow 0}\left\{\Delta\sum_{n\geq 0}\left[\sup_{n\Delta\leq t\leq (n+1)\Delta} L_{ij}(t) - \inf_{n\Delta\leq t\leq (n+1)\Delta} L_{ij}(t)\right]\right\} = 0,$$

Equation (6.8) has a unique solution $\Theta = \psi * L(t)$ belonging to $\mathbf{B}$, and further

$$\lim_{t\to\infty}\Theta_{ij}(t) = \frac{1}{m}\sum_{\ell\in E} v_\ell \int_0^\infty L_{\ell j}(t)dt. \tag{6.9}$$

The following result is an important application of the above theorem.

Proposition 6.1 *The transition function* $P(t) = (P_{(t)})$ satisfies the following MREij

$$P(t) = I(t) - H(t) + Q * P(t),$$

which, under assumptions of Theorem 6.2, has the unique solution

$$P(t) = \psi * (I - H(t)), \tag{6.10}$$

and, also from Theorem 6.2, for any $i, j \in E,$

$$\lim_{t \to \infty} P_{ji}(t) = v_i m_i / m =: \pi_i. \tag{6.11}$$

Here, $H(t) = diag\,(H_i(t))$ *is a diagonal matrix.*

It is worth noticing that, in general, the stationary distribution π of the semi-Markov process Z is not equal to the stationary distribution v of the embedded Markov chain (J_n). Nevertheless, we have $\pi = v$ when, for example, m_i is independent of $i \in E$.

Example Alternating Renewal Process

The transition function of the SMP Z is

$$P(t) = M * \begin{pmatrix} 1 - F & F * (1 - G) \\ G * (1 - F) & 1 - G \end{pmatrix}(t)$$

where M is the renewal function of the distribution function $F*G$, i.e.,

$$M(t) = \sum_{n=0}^{\infty} (F * G)^{(n)}(t).$$

Finally, by the Markov renewal theorem, we get also

$$\lim_{t \to \infty} P(t) = \begin{pmatrix} m_1 & m_0 \\ m_1 & m_0 \end{pmatrix} \Big/ (m_1 + m_0),$$

where m_0 and m_1 are the mean values of F and G, respectively. So, the limiting probability of the semi-Markov process is

$$\pi_1 = \frac{m_1}{m_1 + m_0}, \quad \pi_0 = \frac{m_0}{m_1 + m_0}.$$

6.2.3 Nonparametric Estimation

Statistical inference for semi-Markov processes is provided in several papers. Moore and Pyke [47] studied empirical estimators for finite semi-Markov kernels; Lagakos, Sommer, and Zelen [37] gave maximum likelihood estimators for nonergodic finite semi-Markov kernels; Akritas and Roussas [2] gave parametric local asymptotic normality results for semi-Markov processes; Gill [20] studied Kaplan–Meier type estimators by point process theory; Greenwood and Wefelmeyer [22] studied efficiency of empirical estimators for linear functionals in the case of a general state space; Ouhbi and Limnios [49] studied nonparametric estimators of semi-Markov kernels, nonlinear functionals of semi-Markov kernels, including Markov renewal matrices and reliability functions [50], and rate of occurrence of failure functions [51]. Also see the book [1]. We will give here some elements of the nonparametric estimation of semi-Markov kernels.

Let us consider an observation of an irreducible semi-Markov process Z, with finite state space E, up to a fixed time T, i.e., $(Z_s, 0 \leq s \leq T) \equiv (J_0, J_1, \ldots, J_{N(T)};$ $X_1, \ldots, X_{N(T)-1}, T - S_{N(T)})$, if $N(T) > 0$ and $(Z_s, 0 \leq s \leq T) \equiv (J_0)$ if $N(T) = 0$.

The empirical estimator $\widehat{Q}_{ij}(t, T)$ of $Q_{ij}(t)$ is defined by

$$\widehat{Q}_{ij}(t, T) = \frac{1}{N_i(T)} \sum_{k=1}^{N(T)} \mathbf{1}_{\{J_{k-1}=i, J_k=j, X_k \leq t\}}. \tag{6.12}$$

We denote by $\xrightarrow{a.s.}$ and $\xrightarrow{D}$ the almost sure convergence and convergence in distribution, respectively. Denote by $N(a, b)$ the normal random variable with mean a and variance b. Then, we have the following result from Moore and Pyke [47], see also [51] and [38] for a functional version.

Theorem 6.3 ([47]) *For any fixed $i, j \in E$, as $T \rightarrow \infty$, we have:*

(a) *(Strong consistency)* $\max_{i,j} \sup_{t \in (0,T)} \left| \widehat{Q}_{ij}(t, T) - Q_{ij}(t) \right| \xrightarrow{a.s.} 0,$

(b) *(Asymptotic normality)* $T^{1/2} \left(\widehat{Q}_{ij}(t, T) - Q_{ij}(t) \right) \xrightarrow{D} N(0, \sigma_{ij}^2(t))$, *where* $\sigma_{ij}^2(t) :$
$= \mu_{ii} Q_{ij}(t) \left[1 - Q_{ij}(t) \right].$

We define estimators of the Markov renewal function and of transition probabilities by *plug in* procedure, i.e., replacing semi-Markov kernel $Q_{ij}(t)$ in Eqs. (6.6) and (6.10) by the empirical estimator kernel $\widehat{Q}_{ij}(t, T)$.

The following two results concerning the Markov renewal function estimator, $\widehat{\psi}_{ij}(t, T)$, and the transition probability function estimator, $\widehat{P}_{ij}(t, T)$, of Z were proved by Ouhbi and Limnios [51].

Theorem 6.4 ([51]) *The estimator $\widehat{\psi}_{ij}(t, T)$ of the Markov renewal function $\psi_{ij}(t)$ satisfies the following two properties:*

(a) *(Strong consistency) it is uniformly strongly consistent, i.e., as $T \rightarrow \infty$,*

$$\max_{i,j} \sup_{t \in (0,T)} \left| \widehat{\psi}_{ij}(t, T) - \psi_{ij}(t) \right| \xrightarrow{a.s.} 0.$$

(b) (*Asymptotic normality*) *For any fixed* $t > 0$, *it converges in distribution, as* $T \to \infty$, *to a normal random variable, i.e.*,

$$T^{1/2}(\widehat{\psi}_{ij}(t,T) - \psi_{ij}(t)) \xrightarrow{D} N(0, \sigma_{ij}^2(t)),$$

where $\sigma_{ij}^2(t) = \sum_{r \in E} \sum_{k \in E} \mu_{rr}\{(\psi_{ir} * \psi_{kj})^2 * Q_{rk} - (\psi_{ir} * \psi_{kj} * Q_{rk})^2\}(t)$.

Theorem 6.5 ([49]) *The estimator* $\widehat{P}_{ij}(t,T)$ *of the transition function* $P_{ij}(t)$, *satisfies the following two properties*:

(a) (*Strong consistency*) *for any fixed* $L > 0$, *we have, as* $T \to \infty$

$$\max_{i,j} \sup_{t \in [0,L]} \left| \widehat{P}_{ij}(t,T) - P_{ij}(t) \right| \xrightarrow{a.s.} 0;$$

(b) (*Asymptotic normality*) *For any fixed* $t > 0$, *we have, as* $T \to \infty$,

$$T^{1/2}(\widehat{P}_{ij}(t,T) - P_{ij}(t)) \xrightarrow{D} N(0, \sigma_{ij}^2(t)),$$

where

$$\sigma_{ij}^2(t) = \sum_{r \in E} \sum_{k \in E} \mu_{rr}[(1 - H_i) * B_{irkj} - \psi_{ij} \mathbf{1}_{\{r=j\}}]^2 * Q_{rk}(t)$$
$$- \{[(1 - H_i) * B_{irkj} - \psi_{ij} \mathbf{1}_{\{r=j\}}] * Q_{rk}(t)\}^2,$$

and

$$B_{irkj}(t) = \sum_{n=1}^{\infty} \sum_{\ell=1}^{n} Q_{ir}^{(\ell-1)} * Q_{kj}^{(n-\ell)}(t).$$

6.3 Reliability of Semi-Markov Systems

6.3.1 Reliability Modeling

For a stochastic system with state space E described by a semi-Markov process Z, let us consider a partition U, D of E, i.e., $E = U \cup D$, with $U \cap D = \emptyset$, $U \neq \emptyset$, and $D \neq \emptyset$. The set U contains the up states and D contains the down states of the system. The reliability is defined by $R(t) := P(Z_s \in U, \forall s \in [0, t])$. For the finite state space case, without loss of generality, let us enumerate first the up states and next the down states, i.e., for $E = \{1, 2, ..., d\}$, we have $U = \{1, ..., r\}$ and $D = \{r + 1, ..., d\}$.

Reliability. Define the hitting time T of D, that is,

$$T = \inf\{t \geq 0 : Z_t \in D\}, \quad (\inf \emptyset = +\infty),$$

and the conditional reliability

$$R_i(t) = P_i(T > t) = P_i(Z_s \in U, \forall s \le t), \quad i \in U.$$

Of course, $R_i(t) = 0$ for any $i \in D$ and $t \ge 0$.

Proposition 6.2 *The conditional reliability function $R_x(t)$ satisfies the Markov renewal equation (MRE):*

$$R_i(t) = \overline{H}_i(t) + \sum_{j \in U} \int_0^t Q_{ij}(ds) R_j(t - s), \quad i \in U. \tag{6.13}$$

Hence, under assumption A0 (corresponding to U set), the reliability is given by the unique solution of the above equation, and the (unconditional) reliability is

$$R(t) = \sum_{i \in U} \sum_{j \in U} \int_0^t \alpha_i \psi_{ij}(ds) \overline{H}_j(t - s), \tag{6.14}$$

or, equivalently, in matrix form

$$R(t) = \alpha_0 (I - Q_0((t)))^{(-1)} * \overline{H}_0(t), \tag{6.15}$$

where ψ is the Markov renewal function, corresponding to the sub semi-Markov kernel Q_0, which is the restriction of the semi-Markov kernel Q on $U \times U$, and $t \ge 0$; $\overline{H}_0(t) = diag(\overline{H}_i(t); i \in U)$ a diagonal matrix; and $\alpha = (\alpha_i, i \in E)$ the initial distribution of Z.

The reliability function is defined for both kind of systems: nonrepairable (states U are transient) and repairable (the system is ergodic or equivalently Assumptions A0–A1 hold). In the first case, T denotes the absorption time into the set D, and in the former one, it denotes the (first) hitting time of the set D.

The above formula (6.15) is a generalization of the phase-type distribution functions.

Availability. As we have seen in Chap. 2 , we have several types of availability. In the present chapter, we will adapt the general definition to the case of semi-Markov processes.

Instantaneous or Point availability This is the availability of a system at time $t \ge 0$. Define the conditional point availability by

$$A_i(t) = P_i(Z_t \in U), \quad i \in E,$$

where π is the stationary probability of Z. From the Markov renewal theorem, we have $\pi_i = v_i m_i / m$ (see Proposition 6.1).

Proposition 6.3 *The conditional availability function $A_i(t)$ satisfies the MRE:*

$$A_i(t) = \mathbf{1}_U(i)\overline{H}_i(t) + \sum_{j \in E} \int_0^t Q_{ij}(ds)A_j(t - s).$$

Hence, under assumption A0, the availability is given by the unique solution of the above equation, and then the (unconditional) availability is

$$A(t) = \sum_{i \in E}\sum_{j \in U} \int_0^t \alpha_i \psi_{ij}(ds)\overline{F}_j(t - s), \tag{6.16}$$

or, equivalently, in matrix form

$$A(t) = \alpha(I - Q((t)))^{(-1)} * \overline{H}(t), \tag{6.17}$$

Since $\{Z_s \in U, 0 \leq s \leq t\} \subset \{Z_t \in U\}$, over $\{Z_0 \in U\}$, we obtain the important inequality $R(t) \leq A(t)$, for all $t \geq 0$. In case where all states U are transient, then formula (6.16) can be reduced to (6.15), and then, we have $A(t) = R(t)$ for all $t \geq 0$.

Steady-state availability. This is the pointwise availability under stationary process Z, that is

$$A = P_\pi(Z_t \in U).$$

Of course, the above probability is independent of t.

Again from the Markov renewal theorem, applied to (6.16), it is easy to see that $\lim_{t \to \infty} A_i(t) = A_s$.

Proposition 6.4 *Under assumptions A0–A1, the steady state availability is*

$$A_s = \frac{1}{m}\sum_{i \in E} v_i m_i.$$

Average availability. The average availability over $(0, t]$ is defined by

$$A_{av}(t) := \frac{1}{t}\int_0^t A(s)ds. \tag{6.18}$$

It is clear that under assumptions A0–A1, we have $\lim_{t \to \infty} A_{av}(t) = A$.

Interval availability. This is the probability that system is in the set of up states U at time t, and it will remain there during the time interval $[t, t + s]$, denoted by

$A(t, s)$ $(t \geq 0, s \geq 0)$, that is $A(t, s) := P(Z_u \in U, \forall u \in [t, t + s])$. This generalize the reliability and availability functions, since: $A(0,t) = R(t)$ and $A(t, 0) = A(t)$, (see e.g., [5]). Denote by $A_i(t, s)$ the conditional interval availability on the event $\{Z_0 = i\}$.

Proposition 6.5 *The interval availability function $A_i(t, s)$ satisfies the MRE:*

$$A_i(t,s) = \mathbf{1}_U(i)\overline{H}_i(t+s) + \sum_{j \in U} \int_0^{t+s} Q_{ij}(du)A_j(t-u,s).$$

Hence, under assumption A0, the interval availability is given by the unique solution of the above equation, and the (unconditional) interval availability is

$$A(t,s) = \sum_{i \in E} \sum_{j \in U} \int_0^{t+s} \alpha_i \psi_{ij}(du)\overline{H}_j(t+s-u). \tag{6.19}$$

And under the additional assumption A1, the limiting interval availability, for fixed $s > 0$, is given by

$$A_\infty(s) := \lim_{t \to \infty} A_i(t, s) = \frac{1}{m} \sum_{j \in U} v_j \int_0^\infty \overline{H}_j(t+s)dt.$$

Mean times.   The mean times presented here play an important role in reliability practice.

MTTF: Mean Time To Failure.   Let us define the conditional mean time to failure $MTTF_i = E_i[T]$, $i \in U$, with values in $[0, \infty]$, and the (unconditional) mean time to failure $MTTF = E[T]$. Of course, $MTTF_i = 0$, for any $i \in D$.

Proposition 6.6 ([43]) *The $MTTF_i$ is given by*

$$MTTF = \alpha_0(I - P)^{-1}m_0,$$

where α_0 and m_0 are the restriction of the row vector α of the initial law and of the column vector m of the mean sojourn times, on the up states U, respectively.

MUT: Mean Up Time. This time concerns the mean up time under condition that the underling semi-Markov process is in steady state. We obtain this mean time by calculating first the entry distribution to U set under stationary process assumption:

Proposition 6.7 ([43]) *Under assumptions A0–A1, the entry distribution in U is*

$$\beta(j) = \lim_{t \to \infty} P(Z_t \in B \mid Z_{t-} \in D, Z_t \in U)$$

$$= P_v(J_{n+1} \in B \mid J_n \in D, J_{n+1} \in U)$$

$$= \frac{\sum_{i \in D} v_i P(i,j)}{\sum_{\ell \in U} \sum_i v_i P(i,\ell)}$$

Consequently, we have the following result.

Proposition 6.8 ([43]) *Under assumptions A0–A1, the MUT is:*

$$MUT = \frac{v_0 m_0}{v_1 P_{10} 1_r}$$

where v_0, m_0 are the restrictions of the vectors v (d-dimensional row vector) and μ(d-dimensional column vector), respectively, to the U-set; P_{10} is the restriction of the matrix P on the $D \times U$-set; and 1_r is the column r-dimensional vector of ones $(1, \ldots, 1)$.

Rate of occurrence of failures (ROCOF). The rate of occurrence of failure has been already defined in Chap. 3. The failure counting point process considered here gives the number of times the semi-Markov process Z has visited set D in the time interval $(0, t]$, that is,

$$N_F(t) := \sum_{s \leq t} \mathbf{1}_{\{Z_{t-} \in U, Z_t \in D\}}.$$

It is worth noticing that the above summation is meaningful since in the whole paper the process Z is supposed to be regular (Assumption A0) and so, in every finite interval $(0, t]$ the number of jumps of the process Z is finite. The results here are generalization of those obtained in [51] for the finite state space case. The proof here is new, since the proof given in the finite case does not apply here.

The intensity of the process $N_F(t)$, $t \geq 0$, or the rate of occurrence of failure, $m(t)$, is defined by

$$m(t) := \frac{d}{dt} EN_F(t).$$

The following additional assumptions will be used in the sequel.

Assumptions 6.2
A2: $\|F\|_{\infty,\Delta} = O(\Delta)$, for $\Delta \downarrow 0$;
A3: the semi-Markov kernel Q is absolutely continuous with respect to Lebesgue measure on $\mathbf{R}_+$, with Radon–Nikodym derivative q, that is $Q_{ij}(dt) = q_{ij}(t)dt$, for all $i, j \in E$;
A4: the function $q_{ij}(\cdot)$ is direct Riemann integrable, for any $i, j \in E$ (see e.g., [4, 33, 60]).

Let us also define the hazard rate function $\lambda(x, t)$ of the holding time distribution in state $i \in E$, $F_i(t)$, that is,

$$\lambda_i(t) = \frac{1}{\overline{F}_i(t)} \frac{d}{dt} F_i(t), \quad t \geq 0.$$

A5: The function $\lambda_i(t)$ exists for every $i \in E$ and belongs to the space $\mathbf{B}_1$.

It is worth noticing that assumption A2 implies A0 and assumption A3 implies that λ exists.

Proposition 6.9 ([51]) *Under Assumptions A2, A3, and A5, we have*

$$m(t) = \sum_{\in E} \sum_{j \in U} \sum_{m \in D} \int_0^t \alpha_i \psi_{ij}(du) q_{jm}(t - y),$$

and under the additional assumption A4, the asymptotic rate of occurrence is

$$m_\infty := \lim_{t \to \infty} m(t) = \sum_{i \in U} \sum_{j \in D} v_i P(i, j) / m.$$

Residual lifetime.   The residual lifetime at time $t \geq 0$ is the time left up to the failure for a component starting its life at time 0 and it is still alive at time t.

The distribution of the residual lifetime is defined by

$$F_t(s) = P(T \leq t + s \mid T > t) = 1 - \frac{R(t + s)}{R(t)}.$$

The mean residual lifetime is

$$E[T - t \mid T > t] = \int_t^\infty R(u) \frac{du}{R(t)}.$$

Hence, in the semi-Markov setting, we have

$$F_t(s) = 1 - \frac{\alpha_0(\psi_0 * \overline{F})(t + s)}{\alpha_0(\psi_0 * \overline{F})(t)}.$$

Example A Three-State Semi-Markov System

Let us consider a three-state semi-Markov system $E = \{1, 2, 3\}$ with semi-Markov kernel given by the following data. Let be $Q_{12}(t) = 1 - \exp(-0.1t), Q_{21}(t) = \frac{2}{3}[1 - \exp(-t^{1.5})]$ and $Q_{23}(t) = \frac{1}{3}[1 - \exp(-t^{0.9})]$. The other elements of the semi-Markov kernel are identically null. We suppose that the up states are $U = \{1, 2\}$

Fig. 6.2 Reliability of the three state system given in example

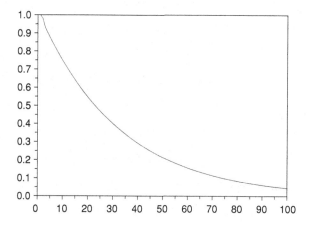

and $D = \{3\}$ is the down state and that system starts in state 1 at time $t = 0$. So, the reliability is given by

$$R(t) = (1 - Q_{12} * Q_{21})^{(-1)} * (1 - Q_{12} * Q_{23} - Q_{12} * Q_{21})(t),$$

where $(1 - Q_{12}{}^*Q_{21})^{(-1)}(t) = \sum_{n \geq 0}(Q_{12}{}^*Q_{21})^{(n)}(t)$. Figure 6.2 gives the reliability of this system. It is worth noticing that since the system is absorbing, we have $A(t) = R(t)$, for all $t \geq 0$.

6.3.2 Reliability Estimation

From estimator (6.12), the following *plug in* estimator of reliability is proposed

$$\widehat{R}(t, T) = \widehat{\alpha}_0(I - \widehat{Q}_0)^{(-1)} * \widehat{\overline{H}}_0(t, T).$$

It is worth noticing that since we have one trajectory the estimator $\widehat{\alpha}_0$ of the initial distribution is given by $\widehat{\alpha}_i = 1_i(Z_0), i \in E$. The following properties are fulfilled by the above reliability estimator.

Theorem 6.6 ([50]) *For any fixed $t > 0$ and for any $L \in (0, \infty)$, we have*

(a) (*Strong consistency*)

$$\sup_{0 \leq t \leq L} | \widehat{R}(t, T) - R(t) | \xrightarrow{a.s.} 0, \quad as \quad T \to \infty,$$

(b) (*Asymptotic normality*)

$$T^{1/2}(\widehat{R}(t, T) - R(t)) \xrightarrow{D} N(0, \sigma_R^2(t)), \quad as \quad T \to \infty,$$

where

$$\sigma_R^2(t) = \sum_{i \in U} \sum_{j \in E} \mu_{ii} \{ (B_{ij} \mathbf{1}_{\{j \in U\}} - \sum_{r \in U} \alpha(r) \psi_{ri})^2 * Q_{ij}(t)$$

$$- [(B_{ij} \mathbf{1}_{\{j \in U\}} - \sum_{r \in U} \alpha(r) \psi_{ri}) * Q_{ij}(t)]^2 \}$$

and

$$B_{ij}(t) = \sum_{n \in E} \sum_{k \in U} \alpha(i) \psi_{ni} * \psi_{jk} * (I - H_k(t))).$$

Availability estimation.

$$\widehat{A}(t, T) = \widehat{\alpha}(I - \widehat{Q})^{(-1)} * \widehat{\overline{H}}(t, T).$$

Theorem 6.7 ([50]) *For any fixed $t > 0$, and $L \in (0, \infty)$, we have*

(a) *(Strong consistency)*

$$\sup_{0 \le t \le L} | \widehat{A}(t, T) - A(t) | \overset{a.s.}{\to} 0, \quad as \quad T \to \infty,$$

where $L \in \mathbf{R}_+$.
(b) *(Asymptotic normality)*

$$T^{1/2}(\widehat{A}(t, T) - A(t)) \overset{D}{\to} N(0, \sigma_A^2(t)),$$

as $T \to \infty$, where

$$\sigma_A^2(t) = \sum_{i \in U} \sum_{j \in E} \mu_{ii} \{ (B_{ij} \mathbf{1}_{\{j \in U\}} - \sum_{r \in U} \alpha(r) \psi_{ri})^2 * Q_{ij}(t)$$

$$- [(B_{ij} \mathbf{1}_{\{j \in U\}} - \sum_{r \in U} \alpha(r) \psi_{ri}) * Q_{ij}(t)]^2 \}$$

$$+ \sum_{i \in D} \sum_{j \in E} \mu_{ii} \{ B_{ij}^2 * Q_{ij}(t) - [B_{ij} * Q_{ij}(t)]^2 \}.$$

Estimation of MTTF.

$$\widehat{MTTF}_T = \widehat{\alpha}_0 (I - \widehat{P}_0(T))^{-1} \widehat{m}_0(T), \tag{6.20}$$

where the estimators $\widehat{P}_0(T) = (\widehat{p}_{ij}(T), i, j \in U)$, and $\widehat{m}_0(T) = (\widehat{m}_i(T), i \in U)$ are defined as follows:

$$\widehat{p}_{ij}(T) = \frac{N_{ij}(T)}{N_i(T)}, \quad \widehat{m}_i(T) := \int_0^\infty \widehat{S}_i(t; T) dt = \frac{1}{N_i(T)} \sum_{l=1}^{N_i(T)} X_{il},$$

for $i, j \in U$, where X_{il} is the l-th sojourn time in the state i, and $\widehat{S}_i(t; T)$, the empirical estimator of the survival function $S_i(t)$ of sojourn time in state i. In the case where $N_i(T) = 0$, we set $\widehat{S}_i(t; T) = 0$.

Properties of MTTF Estimator.

Theorem 6.8 ([45])

(a) (*Strong consistency*) *The estimator of the MTTF is consistent in the sense that*

$$\widehat{MTTF}_T \longrightarrow MTTF \quad P_i - a.s., \quad \text{as} \quad T \to \infty,$$

for any $i \in E$.

(b) (*Asymptotic normality*) *The r.v.* $\sqrt{T}(\widehat{MTTF}_T - MTTF)$ *converges in distribution to a normal random variable with mean zero and variance*

$$\sigma^2 = \sum_{i \in U} \sum_{j \in U} \alpha_i^2 m_i^2 \left[\sum_{k \in U} A_{ik}^2 p_{kj}(1 - p_{kj}) + \sum_{k \in U} A_{kj}^2 p_{ik}(1 - p_{ik}) \right.$$
$$\left. + \sum_{l \in U} \sum_{r \in U} A_{il}^2 A_{rj}^2 p_{lr}(1 - p_{lr}) + A_{ij}^2 \sigma_j^2 \right],$$

where $\sigma_j^2 := m_j^{(2)} - m_j^2$ *is the variance of* X_{j1}, $m_j^{(2)} = \int_0^\infty t^2 dF_j(t)$, *and* $A_{ij} = (I - P_0)^{-1}_{ij}$, $i, j \in U$.

Estimation of the system failure rate.

The failure rate of the semi-Markov system is defined as follows

$$\lambda(t) := \lim_{h \downarrow 0} \frac{1}{h} P(Z_{t+h} \in D \mid Z_u \in U, \forall u \le t). \tag{6.21}$$

From this definition, we get:

$$\lambda(t) = \frac{\alpha_0 \psi_0 * H_0'(t) \mathbf{1}}{\alpha_0 \psi_0 * (I - H_0(t)) \mathbf{1}}, \tag{6.22}$$

where $H_0'(t)$ is the diagonal matrix of derivatives of $H_i(t)$, i.e., $H_0'(t) := diag (H_i'(t), i \in U)$.

Replacing Q, ψ, H by their estimator in relation (6.22), we get an empirical estimator of the system failure rate λ at time t, i.e.,

$$\widehat{\lambda}(t, T) = \frac{\widehat{\alpha}_0 \widehat{\psi}_0 * \widehat{H}_0'(t, T) \mathbf{1}}{\widehat{\alpha}_0 \widehat{\psi}_0 * (I - \widehat{H}_0(t, T)) \mathbf{1}}, \tag{6.23}$$

where the derivative $\widehat{H}'(t, T)$, en t, is estimated by

$$\frac{\widehat{H}(t + \Delta, T) - \widehat{H}(t, T)}{\Delta}.$$

Theorem 6.9 ([49]) *The estimator of the failure rate given by (6.23) satisfy the following properties:*

(a) *(Strong consistency) If the semi-Markov kernel is continuously differentiable, then the estimator is uniformly strongly consistent, i.e., for any fixed $L \in (0, \infty)$,*

$$\sup_{0 \leq t \leq L} \left| \widehat{\lambda}(t, T) - \lambda(t) \right| \xrightarrow{a.s.} 0, \quad T \to \infty.$$

(b) *(Asymptotic normality) If f_{ij} is twice continuously differentiable for all $i, j \in E$, then*

$$T^{\frac{1-\delta}{2}}(\widehat{\lambda}(t, T) - \lambda(t)) \xrightarrow{D} N(0, \sigma_\lambda^2), \quad T \to \infty,$$

where $0 < \delta < 1/2$, and

$$\sigma_\lambda^2 = \frac{1}{(R(t))^2} \sum_{j \in U} \mu_{jj} \left\{ [(\sum_{i \in U} \alpha(i) \psi'_{0;ij})^2 * H_j](t) - (\sum_{i \in U} \alpha(i) [\psi'_{0;ij} * H_j](t))^2 \right\}.$$

By Proposition 6.9, we propose the following estimator for the ROCOF of the semi-Markov system

$$\widehat{ro}(t, T) = \sum_{i \in U} \sum_{j \in D} \sum_{l \in E} \alpha(l) \widehat{\psi}_{li} * \widehat{q}_{ij}(t, T). \tag{6.24}$$

The following theorem gives the asymptotic properties of the ROCOF estimator.

Theorem 6.10 ([51])

(a) *(Strong consistency) Under assumptions of Proposition 6.9, the estimator (6.24) is uniformly strongly consistent, i.e., for any fixed $L \in (0, \infty)$,*

$$\sup_{t \in [0,L]} |\widehat{ro}(t, T) - m(t)| \xrightarrow{a.s.} 0, \quad T \to \infty.$$

(b) *(Asymptotic normality) Under Assumptions A2–A5, and if further $q_{ij}(.)$ is twice continuously differentiable in t for any $i \in U$ and $j \in D$, then the estimator $\widehat{ro}(t, T)$ is asymptotically normal with mean $m(t)$ and variance*

$$\sigma^2(t) := \sum_{i \in U} \sum_{j \in D} \mu_{ii} \frac{\sum_{l \in E} \alpha(l) \cdot \psi_{li} * q_{ij}(t)}{T^{1-\delta}} + O(T^{-1}),$$

where μ_{ii} is the mean recurrence time between two successive visits to state i and $0 < \delta < 1$.

Fig. 6.3 A three state semi-Markov system

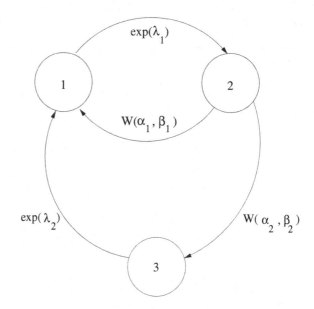

For further results on estimation, see [44, 45, 50, 51, 52].

Example Estimation of A Three-state Space System

Let us consider a three-state system (see Fig. 6.3). The conditional transition functions are indicated in the same figure. We have two exponentials with parameters λ_1 and λ_2 and two Weibull with parameters $(\alpha_1; \beta_1)$ and $(\alpha_2; \beta_2)$. The distribution function of a two parameter $(\alpha; \beta)$ Weibull distribution is $F(t) = 1 - \exp(-(t/\alpha)^\beta$, for $t \geq 0$.

Then, the semi-Markov kernel is

$$Q(t) = \begin{pmatrix} 0 & Exp(\lambda_1) & 0 \\ pW(\alpha_1, \beta_1) & 0 & (1-p)W(\alpha_2, \beta_2) \\ Exp(\lambda_2) & 0 & 0 \end{pmatrix}$$

The transition probability matrix of the embedded Markov chain J is

$$P = \begin{pmatrix} 0 & 1 & 0 \\ p & 0 & 1-p \\ 1 & 0 & 0 \end{pmatrix}$$

Figure 6.4 gives the reliability and availability estimations, and Fig. 6.5. gives the confidence intervals of reliability in levels 90 and 80%.

Fig. 6.4 Reliability and
availability estimation of the
three state semi-Markov
system

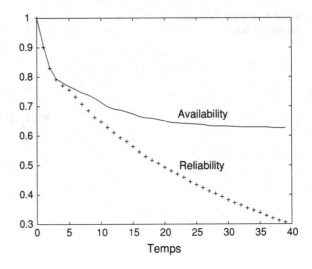

Fig. 6.5 Reliability
confidence interval
estimation at levels 90
and 80% of the three state
semi-Markov system

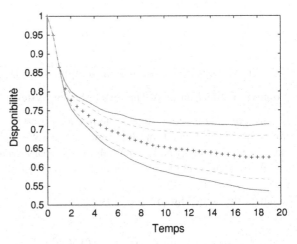

6.4 Reliability in Discrete Time

6.4.1 Semi-Markov Chains

Let us consider a Markov renewal process (J_k, S_k) in discrete time $k \in \mathbf{N}$, with
state space E at most countable as in the previous section. The semi-Markov kernel
q is defined by

$$q_{ij}(k) := P(J_{n+1} = j, S_{n+1} - S_n = k \mid J_n = i),$$

where $i, j \in E$, and $k, n \in \mathbf{N}$.[6, 62]

The process (J_n) is the embedded Markov chain of the MRP (J_n, S_n) with transition matrix P. The semi-Markov kernel q can be written as

$$q_{ij}(k) = p_{ij}f_{ij}(k),$$

where $f_{ij}(k) := P(S_{n+1} - S_n = k \mid J_n = i, J_{n+1} = j)$, the conditional distribution of the sojourn time, and $P = (p_{ij})$ is the transition probability matrix of the embedded Markov chain (J).

Consider a function $\varphi : E \times \mathbf{N} \to \mathbf{R}$, and define its convolution by q as follows

$$q * \varphi(i, k) = \sum_{j \in E} \sum_{\ell=0}^{k} q_{ij}(\ell) \varphi(j, k - \ell).$$

Define now the n-fold convolution of q by it self, as follows:

$$q_{ij}^{(n)}(k) = \sum_{r \in E} \sum_{\ell=0}^{k} q_{ir}(\ell) q_{rj}^{(n-1)}(k - \ell), \quad n \geq 2,$$

further $q_{ij}^{(1)}(k) = q_{ij}(k)$, and $q_{ij}^{(0)}(k) = \delta_{ij}\mathbf{1}_{\mathbf{N}}(k)$.

Define also the Markov renewal function ψ by

$$\psi_{ij}(k) := \sum_{n=0}^{k} q_{ij}^{(n)}(k). \tag{6.25}$$

Let us consider two functions $u, w : E \times \mathbf{N} \to \mathbf{R}$. The Markov renewal equation is here

$$u(i, k) = w(i, k) + \sum_{r \in E} \sum_{\ell=0}^{k} q_{ir}(\ell) u(r, k - \ell). \tag{6.26}$$

For example, the Markov Renewal function (6.25) can be written in the following form which is a particular MRE, in matrix form

$$\psi(k) = q^{(0)}(k) + q * \psi(k).$$

Let us define the semi-Markov chain $Z_k, k \in \mathbf{N}$ associated with the MRP (J_n, S_n). Define first the counting process of jumps, $N(k) := \max\{n \geq 0 : S_n \leq k\}$, and the semi-Markov chain is defined by

$$Z_k = J_{N(k)}, \quad k \in \mathbf{N}.$$

The mean sojourn time in state $i \in E$ is $m_i = \sum_{k \geq 0} \overline{H}_i(k)$ and $m = \sum_{i \in E} v_i m_i$ where $v := (v_i, i \in E)$ is the stationary distribution of the EMC J.

Theorem 6.11 (*Markov Renewal Theorem in discrete time* [60]) *Under the following assumptions*

(1) *The Markov chain (J_n) is ergodic with stationary distribution v.*
(2) $0 < m < \infty$.
(3) *Function $w(i, k)$ satisfies the following condition*

$$\sum_{i\in E}\sum_{k\geq 0} v_i|w(i,k)| < \infty.$$

Then, Eq. (6.26) has a unique solution given by $\psi * w(i, k)$, and

$$\lim_{k\to\infty}\psi * w(i,k) = \frac{1}{m}\sum_{i\in E}v_i\sum_{k\geq 0}w(i,k).$$

The transition function of semi-Markov chain is defined by

$$P_k(i,j) := P(Z_k = j \mid Z_0 = i), \quad i, j \in E,$$

This function fulfils the following MRE

$$P_k(i,j) = \delta_{ij}\overline{F}_i(k) + \sum_{r\in E}\sum_{\ell=0}^{k} q_{ir}(\ell)P_{k-\ell}(r,j).$$

From the Markov renewal theorem, we get the stationary distribution π of Z, as a limit

$$\lim_{k\to\infty}P_k(i,j) = \frac{1}{m}v_j m_j =: \pi_j.$$

6.4.2 Reliability Modeling

Let us consider a stochastic system whose temporal behavior is described by a semi-Markov chain Z with state space $E = \{1, 2, \ldots, d\}$, semi-Markov kernel q, and the initial probability distribution α. As in the continuous time case, the state space E is split into two parts: U containing the up states and D containing the down states.

Define the hitting time T of D, i.e.,

$$T = \inf\{k \geq 0 : Z_k \in D\}, \quad (\inf \varnothing = +\infty),$$

and the conditional reliability by

$$R_i(k) = P_i(T > k) = P_i(Z_\ell \in U, \forall \ell \leq k), \quad i \in U.$$

Of course, $R_i(k) = 0$ for any $i \in D$.

Let us define the Markov renewal function ψ_0,

$$\psi_{0;ij}(k) = \sum_{n=0}^{k} q_{0;ij}^{(n)}(k)$$

where q_0 is the restriction of the semi-Markov kernel q on $U \times U$, and $i, j \in E$, $k \in \mathbf{N}$.

Proposition 6.10 *The reliability function $R_i(k)$ satisfies the MRE:*

$$R_i(k) = \overline{F}_i(t) + \sum_{j \in U} \sum_{\ell=0}^{k} q_{ij}(\ell) R_j(k - \ell), \quad i \in U. \tag{6.27}$$

Hence, the reliability is given by the unique solution of the above equation by

$$R_i(k) = \sum_{j \in U} \psi_{0;ij} * \overline{H}_j(k).$$

Finally, the (unconditional) reliability is given by

$$R(k) = \sum_{i \in U} \sum_{j \in U} \sum_{\ell=0}^{k} \alpha_i \psi_{0;ij}(\ell) \overline{F}_j(k - \ell).$$

Point availability. This is the availability of a system at point time $k \in \mathbf{N}$.

Define the conditional point availability

$$A_i(k) = \mathrm{P}_i(Z_k \in U), \quad i \in E.$$

Proposition 6.11 *The availability function $A_i(k)$ satisfies the MRE:*

$$A_i(k) = \mathbf{1}_U(i) \overline{F}_i(k) + \sum_{j \in E} \sum_{\ell=0}^{k} Q_{ij}(\ell) A_j(k - \ell).$$

Hence, the unconditional availability is given by

$$A(k) = \sum_{i \in E} \sum_{j \in U} \sum_{\ell=0}^{k} \alpha_i \psi_{ij}(\ell) \overline{F}_j(k - \ell). \tag{6.28}$$

Steady-state availability. This is the pointwise availability under stationary process Z. That is

$$A = \mathrm{P}_\pi(Z_k \in U) = \sum_{i \in U} v_i m_i / m.$$

Of course, the above probability is independent of k, and it is given by the Markov renewal theorem as follows.

From the Markov renewal theorem applied to (6.28), it is easy to see that

$$\lim_{k \to \infty} A_i(k) = A,$$

for any $i \in E$.

Average availability over $[0, k]$. This availability is defined by

$$A_{av}(k) := \frac{1}{k+1} \sum_{\ell=0}^{k} A(\ell).$$

Interval availability The interval availability in discrete time, $A(k,p), k, p \in \mathbf{N}$, as in the continuous time case, is defined by

$$A(k,p) = P(Z_\ell \in U, \forall \ell \in [k, k+p]).$$

Proposition 6.12 *The interval availability function* $A_i(k, p)$ *satisfies the MRE:*

$$A_i(k,p) = \mathbf{1}_U(i)\overline{F}_i(k+p) + \sum_{j \in U} \sum_{\ell=0}^{k+p} Q_{ij}(\ell)A_j(k-\ell,p).$$

   Hence, the interval availability is given by the unique solution of the above equation by

$$A_i(k,p) = (\psi * \overline{F}.\mathbf{1}_U)(x, k+p).$$

The unconditional interval availability is given by

$$A(k,p) = \alpha(\psi * \overline{F}.\mathbf{1}_U)(x, k+p). \tag{6.29}$$

The limiting interval availability, for fixed $p > 0$, *is given by*

$$A_\infty(p) := \lim_{k \to \infty} A_i(k,p) = \frac{1}{m} \sum_{i \in U} \nu_i \sum_{k \geq 0} \overline{F}_i(k+p).$$

   Let us now consider the estimation problem of the semi-Markov kernel and reliability functions given an observation of a semi-Markov process over the integer interval $[0, M]$ as in the continuous time case.

   Let us define the following r.v.

$$N_i(M) := \sum_{k=0}^{N(M)-1} \mathbf{1}_{\{J_k=i\}}$$

$$N_{ij}(M) := \sum_{k=1}^{N(M)} \mathbf{1}_{\{J_{k-1}=i,J_k=j\}}$$

$$N_{ij}(k,M) := \sum_{\ell=1}^{N(M)} \mathbf{1}_{\{J_{\ell-1}=i,J_\ell=j,X_\ell=k\}}.$$

Let us consider the following empirical estimators of p_{ij}, f_{ij}, and q_{ij} for $i, j \in E$ and $i \neq j$:

$$\widehat{p}_{ij}(M) = N_{ij}(M)/N_i(M)$$
$$\widehat{f}_{ij}(k, M) = N_{ij}(k, M)/N_{ij}(M)$$
$$\widehat{q}_{ij}(k, M) = N_{ij}(k, M)/N_i(M)$$

The reliability estimator is

$$\widehat{R}(k, M) = \widehat{\boldsymbol{\alpha}}_1 \left[\widehat{\boldsymbol{\psi}}_{11}(\cdot, M) * \left(\mathbf{I} - \widehat{\mathbf{H}}(\cdot, M)_{11} \right) \right](k)\mathbf{1}_{s_1}$$

The availability estimator is

$$\widehat{A}(k, M) = \widehat{\boldsymbol{\alpha}} \left[\widehat{\boldsymbol{\psi}}(\cdot, M) * \left(\mathbf{I} - \widehat{\mathbf{H}}(\cdot, M) \right) \right](k)\mathbf{1}_{s,s_1}$$

The MTTF estimator is

$$\widehat{MTTF}(M) := \widehat{\boldsymbol{\alpha}}_1(M)(I - \widehat{\mathbf{p}}_{11}(M))^{-1}\widehat{\mathbf{m}}_1(M).$$

Let us now consider the following asymptotic properties for the above estimators.

6.4.2.1 Reliability Estimation

Theorem 6.12 ([6]) *For a discrete-time semi-Markov system, for fixed $k \in \mathbf{N}$:*

(a) *(Strong consistency)* $\widehat{R}(k, M)$ *is strongly consistent, as $M \to \infty$;*
(b) *(Asymptotic normality)*

$$\sqrt{M}[\widehat{R}(k, M) - R(k)] \xrightarrow{D} N(0, \sigma_R^2(k)), \quad M \to \infty.$$

The variance is given by:

$$\sigma_R^2(k) = \sum_{i=1}^{s} \mu_{ii} \left\{ \sum_{j=1}^{s} \left[D_{ij}^U - \mathbf{1}_{\{i \in U\}} \left(\sum_{t \in U} \alpha(t)\Psi_{ti} \right) \right]^2 * q_{ij}(k) \right.$$
$$\left. - \left[\sum_{j=1}^{s} \left(D_{ij}^U * q_{ij} - \mathbf{1}_{\{i \in U\}} \left(\sum_{t \in U} \alpha(t)\psi_{ti} \right) * Q_{ij} \right) \right]^2 (k), \right\}$$

with $D_{ij}^U := \sum_{n \in U} \sum_{r \in U} \alpha(n)\psi_{ni} * \psi_{jr} * (I - diag(Q \cdot \mathbf{1}))_{rr}$.

6.4.2.2 Availability Estimation

Theorem 6.13 ([6]) *For a discrete-time SM system, for fixed $k \in \mathbf{N}$:*

(a) *(Strong consistency)* $\widehat{A}(k,M)$ *is strongly consistent, as $M \to \infty$;*
(b) *(Asymptotic normality)*

$$\sqrt{M}[\widehat{A}(k,M) - A(k)] \xrightarrow{D} N(0, \sigma_A^2(k)), \quad M \to \infty.$$

The variance is given by:

$$\sigma_A^2(k) := \sum_{i=1}^{s} \mu_{ii} \left\{ \sum_{j=1}^{s} \left[D_{ij} - \mathbf{1}_{\{i \in U\}} \left(\sum_{t=1}^{s} \alpha(t) \Psi_{ti} \right) \right]^2 * q_{ij}(k) \right.$$

$$\left. - \left[\sum_{j=1}^{s} \left(D_{ij} * q_{ij} - \mathbf{1}_{\{i \in U\}} \left(\sum_{t=1}^{s} \alpha(t) \psi_{ti} \right) * Q_{ij} \right) \right]^2 (k) \right\}$$

$$D_{ij} := \sum_{n=1}^{s} \sum_{r \in U} \alpha(n) \psi_{ni} * \psi_{jr} * \left(I - diag(Q \cdot \mathbf{1}) \right)_{rr}$$

6.4.2.3 The Failure Rate Estimation

Theorem 6.14 ([6]) *For a discrete-time SM system, for fixed $k \in \mathbf{N}$:*

(a) *(Strong consistency)* $\widehat{\lambda}(k,M)$ *is strongly consistent, as $M \to \infty$;*
(b) *(Asymptotic normality)*

$$\sqrt{M}[\widehat{\lambda}(k,M) - \lambda(k)] \xrightarrow{D} N(0, \sigma_\lambda^2(k)), \quad M \to \infty.$$

The variance is

$$\sigma_\lambda^2(k) = \frac{1}{R^4(k-1)} \sum_{i=1}^{s} \mu_{ii} \left\{ R^2(k) \sum_{j=1}^{s} \left[D_{ij}^U - \mathbf{1}_{\{i \in U\}} \sum_{t \in U} \alpha(t) \Psi_{ti} \right]^2 * q_{ij}(k-1) \right.$$

$$+ R^2(k-1) \sum_{j=1}^{s} \left[D_{ij}^U - \mathbf{1}_{\{i \in U\}} \sum_{t \in U} \alpha(t) \Psi_{ti} \right]^2 * q_{ij}(k) - T_i^2(k)$$

$$+ 2R(k-1)R(k) \sum_{j=1}^{s} \left[\mathbf{1}_{\{i \in U\}} D_{ij}^U \sum_{t \in U} \alpha(t) \Psi_{ti}^+ \right.$$

$$+ \mathbf{1}_{\{i \in U\}} (D_{ij}^U)^+ \sum_{t \in U} \alpha(t) \Psi_{ti} - (D_{ij}^U)^+ D_{ij}^U$$

$$\left. \left. - \mathbf{1}_{\{i \in U\}} \left(\sum_{t \in U} \alpha(t) \Psi_{ti} \right) \left(\sum_{t \in U} \alpha(t) \Psi_{ti}^+ \right) \right] * q_{ij}(k-1) \right\},$$

with

$$T_i(k) := \sum_{j=1}^{s} \Big[R(k)D_{ij}^U * q_{ij}(k-1) - R(k-1)D_{ij}^U * q_{ij}(k)$$
$$- R(k)\mathbf{1}_{\{i \in U\}} \sum_{t \in U} \alpha(t)\psi_{ti} * Q_{ij}(k-1)$$
$$+ R(k-1)\mathbf{1}_{\{i \in U\}} \sum_{t \in U} \alpha(t)\psi_{ti} * Q_{ij}(k) \Big]$$
$$D_{ij}^U := \sum_{n \in U} \sum_{r \in U} \alpha(n)\psi_{ni} * \psi_{jr} * (I - diag(Q \cdot \mathbf{1}))_{rr}$$

$$A^+(k) := A(k+1) \text{ for } A = (A_{ij}(\cdot))_{i,j \in E}$$

6.4.2.4 MTTF Estimation

Theorem 6.15 ([6]) *For a discrete-time SM system*:

(a) (*Strong consistency*) $\widehat{MTTF}(M)$ *is strongly consistent, as* $M \to \infty$;
(b) (*Asymptotic normality*)

$$\sqrt{M}[\widehat{MTTF}(M) - MTTF] \xrightarrow{D} N(0, \sigma_{MTTF}^2), \quad M \to \infty.$$

The variance is

$$\sigma_{MTTF}^2 = \Psi' \Gamma_{11} \Psi'^{\mathsf{T}},$$

where Ψ, Ψ' *and* Γ_{11} *are defined as follows:*

$$\Psi\big((p_{ij})_{i,j \in U}\big) := \alpha_1(I - \mathbf{p}_{11})^{-1}\mathbf{m}_1$$
$$\Psi' := \left(\frac{\partial \Psi}{\partial p_{ij}}\right)_{i,j \in U} -rowvector$$
$$\frac{\partial \Psi}{\partial p_{ij}} = \alpha_1 \frac{\partial(I - \mathbf{p}_{11})^{-1}}{\partial p_{ij}}\mathbf{m}_1 + \alpha_1(I - \mathbf{p}_{11})^{-1}\frac{\partial \mathbf{m}_1}{\partial p_{ij}}$$
$$\frac{\partial(I - \mathbf{p}_{11})^{-1}}{\partial p_{ij}} = (I - \mathbf{p}_{11})^{-1}\frac{\partial \mathbf{p}_{11}}{\partial p_{ij}}(I - \mathbf{p}_{11})^{-1}$$
$$\Gamma_{11} = \begin{pmatrix} \frac{1}{v(1)}\Lambda_1 & 0 & \cdots & 0 \\ 0 & \frac{1}{v(2)}\Lambda_2 & \cdots & 0 \\ \vdots & \vdots & \ddots & \vdots \\ 0 & 0 & \cdots & \frac{1}{v(s_1)}\Lambda_{s_1} \end{pmatrix}$$
$$\Lambda_i = (\delta_{jl}p_{ij} - p_{ij}p_{il})_{j,l \in E}, i \in U$$

6.5 General State Space Semi-Markov Systems

6.5.1 Introduction

Let us consider a semi-Markov process $Z(t)$, $t \geq 0$, with state space the measurable space $(E, \mathscr{E})$, corresponding MRP (J_n, S_n), and semi-Markov kernel Q defined by

$$Q(x, B, t) = P(J_{n+1} \in B, S_{n+1} - S_n \leq t \mid J_n = x),$$

with $x \in E, B \in \mathscr{E}, t \geq 0$, (see e.g., [41, 43]).

The transition kernel of the embedded Markov chain (J_n) is $P(x, B) := Q(x, B, \infty) = \lim_{t \to \infty} Q(x, B, t)$. Let us define the function $F_x(t) := Q(x, E, t)$, which is the distribution function of the holding (sojourn) time in state $x \in E$. The semi-Markov kernel has the following representation

$$Q(x, B, t) = \int_B F(x, y, t) P(x, dy), \quad B \in \mathscr{E}, \tag{6.30}$$

where $F(x, y, t) := P(S_{n+1} - S_n \leq t \mid J_n = x, J_{n+1} = y)$. Define also the r.v. $X_{n+1} := S_{n+1} - S_n$, for $n \geq 0$. We define also the unconditional holding time distribution in state x, by $F_x(t) := Q(x, E, t)$.

Let us consider a real-valued measurable function φ on $E \times \mathbf{R}_+$, and define its convolution by Q as follows

$$(Q * \varphi)(x, t) = \int_E \int_0^t Q(x, dy, ds) \varphi(y, t - s),$$

for $x \in E$, and $t \geq 0$. For two semi-Markov kernels, say Q_1 and Q_2, on $(E, \mathscr{E})$, one defines their convolution by

$$(Q_1 * Q_2)(x, B, t) = \int_E \int_0^t Q_1(x, dy, ds) Q_2(y, B, t - s),$$

where $x \in E, t \in \mathbf{R}_+, B \in \mathscr{E}$. The function $Q_1 * Q_2$ is also a semi-Markov kernel.

We denote $P_{(x,s)}(\cdot) = P(\cdot \mid J_0 = x, S_0 = s)$ and $E_{(x,s)}$, respectively. For a kernel $K(x, B)$, a measure μ, and a point measurable function f, all defined on $(E, \mathscr{E})$, we define:

$$\mu K(B) := \int_E \mu(dx) K(x, B), \quad Kf(x) := \int_E K(x, dy) f(y),$$

and

$$\mu K f = \int_{E \times E} \mu(dx) K(x, dy) f(y).$$

Concerning the rest of the theory we consider here as in Sect. 6.2 except that we replace the symbol of sum over the state space by the symbol of integration over the same space.

6.5.2 Reliability Modeling

In the sequel, we consider a stochastic system of which time behavior is described by a semi-Markov process as defined in the previous section. Consider now a measurable set U in $\mathcal{E}$ containing all up (working) states of the system. The set $E \setminus U$, denoted by D, contains all down (failed) states of the system. In order to avoid trivialities, we suppose that $U \neq \varnothing$ and $U \neq E$. The transition from one state to another state means, physically speaking, the failure or the repair of at least one of the components of the system. The system is operational in U, but no service is delivered if it is in D. Nevertheless, in the case of repairable systems, one repair will return the system from D to U; otherwise, in the case of nonrepairable systems, it will remain in D for ever (see e.g., [43]).

Reliability. Define the hitting time T of D and reliability function as in Sect. 6.2 and we have

Proposition 6.13 ([41]) *The conditional reliability function* $R_x(t)$, $x \in U$ *satisfies the Markov renewal equation (MRE):*

$$R_x(t) = \overline{F}_x(t) + \int_U \int_0^t Q(x, dy, ds) R_y(t - s). \tag{6.31}$$

Hence, under assumption A0 (corresponding to U set), the reliability is given by the unique solution of the above equation, and the (unconditional) reliability is

$$R(t) = \int_U \int_U \int_0^t \alpha(dx) \psi_0(x, dy, ds) \overline{F}_y(t - s), \tag{6.32}$$

where ψ_0 *is the Markov renewal function, corresponding to the sub semi-Markov kernel* Q_0, *which is the restriction of the semi-Markov kernel* Q *on* $U \times \mathcal{E}_U$, *and* $x \in E, B \in \mathcal{E}_U := \mathcal{E} \cap U, t \geq 0$.

The reliability function is defined for both kind of systems: nonrepairable (states U are transient) and repairable (the system is ergodic or equivalently assumption A1 holds). In the first case, T denotes the absorption time into the set D, and in the former one, it denotes the (first) hitting time of the set D.

Availability. In fact, we have several types of availability. The most commonly used ones are the following.

Point availability. This is the availability of a system at time $t \geq 0$. Define the conditional point availability by

$$A_x(t) = P_x(Z_t \in U), \quad x \in E,$$

where π is the stationary probability of Z. From the Markov renewal theorem we have $\pi(dx) = v(dx)m(x)/m$ (see e.g., [43]).

Proposition 6.14 ([41]) *The conditional availability function $A_x(t)$ satisfies the MRE:*

$$A_x(t) = \mathbf{1}_U(x)\overline{F}_x(t) + \int_E \int_0^t Q(x, dy, ds)A_y(t - s).$$

Hence, under assumption A0, the availability is given by the unique solution of the above equation, and the (unconditional) availability is

$$A(t) = \int_E \int_U \int_0^t \alpha(dx)\psi(x, dy, ds)\overline{F}_y(t - s). \tag{6.33}$$

Steady-state availability. This is the pointwise availability under stationary process Z, that is

$$A = P_\pi(Z_t \in U).$$

Of course, the above probability is independent of t.

Again from the Markov renewal theorem, applied to (6.33), it is easy to see that $\lim_{t \to \infty} A_x(t) = A_s$.

Proposition 6.15 ([41]) *Under assumptions A0–A1, the steady state availability is*

$$A = \frac{1}{m} \int_U v(dx)m(x).$$

Average availability. The average availability over $(0, t]$ is defined by

$$A_{av}(t) := \frac{1}{t} \int_0^t A(s)ds. \tag{6.34}$$

It is clear that under assumptions A0–A1, we have $\lim_{t \to \infty} \widetilde{A}(t) = A$.

Interval availability. This is defined as in Sect. 6.3. and we have the following result.

Proposition 6.16 ([41]) *The interval availability function $A_x(t, s)$ satisfies the MRE:*

$$A_x(t, s) = \mathbf{1}_U(x)\overline{F}_x(t + s) + \int\limits_U \int\limits_0^{t+s} Q(x, dy, du)A_y(t - u, s).$$

Hence, under assumption A0, the interval availability is given by the unique solution of the above equation, and the (unconditional) interval availability is

$$A(t, s) = \int\limits_E \int\limits_U \int\limits_0^{t+s} \alpha(dx)\psi(x, dy, du)\overline{F}_y(t + s - u). \tag{6.35}$$

And under the additional assumption A1, the limiting interval availability, for fixed $s > 0$, is given by

$$A_\infty(s) := \lim_{t\to\infty} A_x(t, s) = \frac{1}{m}\int\limits_U v(dy) \int\limits_0^\infty \overline{F}_y(t + s)dt.$$

Mean times.   The mean times presented here play an important role in reliability practice.

MTTF: Mean Time To Failure.   Let us define the conditional mean time to failure function $MTTF_x = E_x[T]$, $x \in U$, with values in $[0, \infty]$, and the (unconditional) mean time to failure $MTTF = E[T]$. Of course, $MTTF_x = 0$, for any $x \in D$.

Proposition 6.17 ([41]) *The $MTTF_x$ is a superharmonic function and satisfies the Poisson equation*

$$(I - P_0)MTTF_x = m(x), \quad x \in U$$

where P_0 is the restriction of P on $U \times \mathscr{E}_U$. Then, we have

$$MTTF_x = G_0 m(x),$$

where G_0 is the potential operator of P_0, that is, $G_0 = \sum_{n \geq 0}(P_0)^n$, (see e.g., [60]), and

$$MTTF = \alpha G_0 m. \tag{6.36}$$

MUT: Mean Up Time.   This time concerns the mean up time under condition that the underling semi-Markov process is in steady state. We obtain this mean time by calculating first the entry distribution to U set under stationary process assumption:

Proposition 6.18 ([41]) *Under assumptions A0–A1, the entry distribution in U is*

$$\beta(B) = \lim_{t \to \infty} P(Z_t \in B \mid Z_{t-} \in D, Z_t \in U)$$

$$= P_v(J_{n+1} \in B \mid J_n \in D, J_{n+1} \in U)$$

$$= \frac{\int_D v(x)P(x,B)}{\int_D v(dx)P(x,U)}.$$

Consequently, we have the following result.

Proposition 6.19 *Under assumptions A0–A1, the MUT is*

$$MUT = \frac{1}{(vP1_U)(U)} \int_U (vP1_D)(dy)G_0m(y).$$

Rate of occurrence of failures (ROCOF). The failure counting point process gives the number of times the semi-Markov process Z has visited set D in the time interval $(0, t]$, that is,

$$N_F(t) := \sum_{s \le t} 1_{\{Z_{t-} \in U, Z_t \in D\}}.$$

It is worth noticing that the above summation is meaningful since in the whole paper the process Z is supposed to be regular. The results here are generalization of those obtained in [51] for the finite state space case. The proof here is new, since the proof given in the finite case does not apply here.

The intensity of the process $N_F(t)$, $t \ge 0$, or the rate of occurrence of failure, $m(t)$, is defined by

$$m(t) := \frac{d}{dt} EN_F(t).$$

Proposition 6.20 ([41]) *Under Assumptions A2, A3, and A5, we have*

$$m(t) = \int_E \int_U \int_D \int_0^t \alpha(dz)\psi(z, dx, du)q(x, dy, t - y),$$

and under the additional assumption A4, the asymptotic rate of occurrence is

$$m_\infty := \lim_{t \to \infty} m(t) = \int_U v(dx)P(x, D)/m.$$

6.6 Stationary Probability Estimation and Availability

6.6.1 Introduction

Let us present an estimation problem of the stationary probability of general state space semi-Markov systems. We will define here the stationary distribution and give an empirical estimator and its (functional) consistence, normality and loglog asymptotic results.

Let us consider a semi-Markov process Z with state space the measurable space $(E, \mathscr{E})$. Let us define the backward and forward recurrence times processes U and V, respectively. Let us consider for a fixed time $t > 0$, the r.v.s

$$U_t := t - S_{N(t)}, \quad \text{and} \quad V_t := S_{N(t)+1} - t,$$

and consider the backward and forward $(E \times \mathbf{R}_+, \mathscr{E} \times \mathscr{B}_+)$-valued processes $(Z_t, U_t, t \geq 0)$ and $(Z_t, V_t, t \geq 0)$, respectively.

Theorem 6.16 *Both processes* $(Z_t, U_t, t \geq 0)$ *and* $(Z_t, V_t, t \geq 0)$ *are Markov processes, with common stationary distribution*

$$\widetilde{\pi}(A \times \Gamma) := \frac{1}{m} \int_A v(\mathrm{d}x) \int_\Gamma [1 - F_x(u)] \mathrm{d}u, \qquad (6.37)$$

on $(E \times \mathbf{R}_+, \mathscr{E} \times \mathscr{B}_+)$.

Both processes are continuous from the right. The semi-group generated by these processes is strongly continuous.

Stationary probability of Z. The marginal law $\pi(A) := \widetilde{\pi}(A \times \mathbf{R}_+)$, on $(E, \mathscr{E})$, is defined to be the stationary probability measure of process Z.

Proposition 6.21 *We have that*

$$\lim_{t \to \infty} \mathrm{P}(Z(t) \in A \mid Z(0) = x) = \pi(A) := \frac{1}{m} \int_A v(\mathrm{d}y) m(y),$$

for any $x \in E$ *and* $A \in \mathscr{E}$.

Example A Continuous State Space Semi-Markov System

Let us consider a semi-Markov process Z with state space $E = [-2, 2]$ and semi-Markov kernel Q given by

$$Q(x, B, t) = \frac{1}{4} \int_B (2 - |y|) \mathrm{d}y (1 - e^{-t|x|+\delta}),$$

where $\delta > 0$, B a measurable subset of E, $x \in E$ and $t \geq 0$.

The stationary probability v of the EMC (J) is given by

$$v(B) = \int_B (2 - |x|)dx/4,$$

and the mean sojourn time $m(x)$ in state x is given by $m(x) = \Gamma(1 + \frac{1}{|x|+\delta})$, where $\Gamma(y) = \int_0^\infty e^{-t}t^{y-1}$, for $y > 0$. So the stationary probability π of the semi-Markov process is

$$\pi(B) = \frac{1}{4m} \int_B (2 - |x|)\Gamma(1 + \frac{1}{|x| + \delta})dx,$$

with

$$m = \frac{1}{4} \int_{-2}^{2} (2 - |x|)\Gamma(1 + \frac{1}{|x| + \delta})dx.$$

The probability $\pi(U)$, where U is the up state space set of the system, is the stationary and asymptotic availability of the system. So, the estimation of π includes as particular result the estimation of stationary availability.

Let us consider a measurable bounded function $g : E \rightarrow \mathbf{R}$. We are going to estimate functionals of the form

$$\tilde{g} := \mathrm{E}_{\tilde{\pi}}g = \int_E \pi(dx)g(x).$$

We propose now the following empirical estimator for $\tilde{v}$:

$$\alpha_T := \frac{1}{T} \int_0^T g(Z(s))ds. \tag{6.38}$$

In the particular case, where $g = \mathbf{1}_A$, with $A \in \mathscr{E}$, we get the following empirical estimator for π,

$$\pi_T(A) := \frac{1}{T} \int_0^T \mathbf{1}_A(Z(s))ds. \tag{6.39}$$

We will present in the sequel the following asymptotic properties of estimator.

- P_x-strong consistency;
- P_x-asymptotic normality;
- $P_{\tilde{\pi}}$-strong invariance principle

$x \in E$.

It is clear that estimators α_T and π_T are $P_{\tilde{\pi}}$-unbiased

$$E_{\tilde{\pi}}\alpha_T = \tilde{g}, \quad \text{and} \quad E_{\tilde{\pi}}\pi_T(A) = \pi(A), \quad T > 0, \quad A \in \mathscr{E}.$$

We have also

$$\alpha_T \xrightarrow{P_{(x,s)} - a.s.} \tilde{v}, \quad \text{as} \quad T \to \infty.$$

Let us consider the family of processes $\alpha_T^\varepsilon, T \geq 0, \varepsilon > 0$,

$$\alpha_T^\varepsilon := \frac{1}{\varepsilon T}\int_0^T g(Z(s/\varepsilon^2))ds - \varepsilon^{-1}\tilde{g} = \frac{1}{\varepsilon T}\int_0^T g_0(Z(s/\varepsilon^2))ds, \qquad (6.40)$$

where $g_0(x) := g(x) - \tilde{g}$.

Consider the associated Markov process $x^0(t)$, $t \geq 0$ to Z, defined by the generator $\mathbf{L}$ as follows

$$\mathbf{L}\varphi(x) = q(x)\int_E P(x,dy)[\varphi(y) - \varphi(x)],$$

where $q(x) = 1/m(x), m(x) = \int_0^\infty \overline{F}_x(t)dt$.

Define the potential operator R_0 of the Markov process by

$$R_0\mathbf{L} = \mathbf{L}R_0 = \Pi - I,$$

where Π is the stationary projector on $\mathbf{B}$, defined by

$$\Pi\varphi(x) = \int_E \pi(dy)\varphi(y)\mathbf{1}_E(x).$$

Assumptions 6.3

IA1: The semi-Markov process Z is uniformly ergodic, that is $\|(P_t - \Pi)\varphi\| \to 0$ as $t \to \infty$ for any $\varphi \in \mathbf{B}$.

IA2: The second moments of the sojourn times are uniformly bounded and

$$m_2(x) = \int_0^\infty t^2 F_x(dt) \leq M < +\infty,$$

and

$$\sup_x \int_T^\infty t^2 F_x(dt) \to 0, \quad T \to \infty.$$

IA3: The embedded Markov chain J is ergodic, with stationary distribution v.

6.6.2 Weak Invariance Principle

The weak invariance principle is stated as follows.

Theorem 6.17 ([33]) *Under Assumptions IA1–A3, the following weak convergence holds*

$$\alpha_t^\varepsilon \Longrightarrow bW(t)/t, \quad \varepsilon \to 0,$$

provided that $b^2 > 0$. The variance coefficient b^2 is

$$b^2 = b_0^2 + b_1,$$

where:

$$b_0^2 = 2 \int_E \pi(dx) a_0(x), \quad a_0(x) = g_0(x) R_0 g_0(x),$$

$$b_1 = \int_E \pi(dx) \mu(x) g_0^2(x), \quad \mu(x) = [m_2(x) - 2m^2(x)]/m(x).$$

Corollary 6.1 *For any $B \in \mathcal{E}$, we have*

$$\varepsilon^{-1}(\pi_t^\varepsilon(B) - \pi(B)) \Longrightarrow bW(t)/t, \quad \varepsilon \to 0,$$

where $\pi_t^\varepsilon := \pi_{t/\varepsilon^2}$, and

$$b^2 = 2 \int_B \pi(dx) R_0(x, B) + (1 - \pi(B))^2 \int_B \pi(dx) \mu(x) + (\pi(B))^2 \int_{E \setminus B} \pi(dx) \mu(x).$$

So, we have, for an observation of a time of order ε^{-2}, the following approximation: $\pi_1^\varepsilon(B) \sim N(\pi(B); \varepsilon^2 b^2)$.

6.6.3 Strong Invariance Principle

Let us present now the strong invariance principle.

Let us consider the semi-Markov kernel $\widetilde{Q}$, defined by $\widetilde{Q}(x, dy \times ds) = P(x, dy) F_y(ds)$, and the transition probability kernel M, whose the iterates are defined as follows

$$M^{(1)}((x, s); B \times [0, t]) = \widetilde{Q}(x, B, t - s)$$

$$M^{(n)}((x, s); B \times [0, t]) = \int_{E \times R_+} M^{(1)}((x, s); dy \times du) M^{(n)}((y, u); B \times [0, t]).$$

Let us define also

$$M^{(n)}g(x,s) = \int\limits_{E \times R_+} M^{(n)}((x,s); dy \times du)g(y,u),$$

where $g(x,s) = g_0(x)s$.

Assumptions 6.4

IA4: The MRP (J_n, X_{n+1}) is stationary, with stationary distribution $\widetilde{v}(dy \times ds) = v(dy)F_y(ds)$.

IA5: We have

$$\sum_{n \geq 1} \left\| M^{(n)}g \right\|_2 < \infty,$$

for the norm $\|f\|_2 := \left(\int_{E \times R_+} f^2 d\widetilde{v} \right)^{1/2}$ on $L^2(\widetilde{v})$.

Theorem 6.18 ([39]) *Let us assume that Assumptions IA1–IA5 hold. If moreover $b^2 > 0$, then the sequence $\left\{ \frac{t(\pi_t^s(B) - \pi(B))}{\varepsilon_n b \sqrt{2 \log \log(\varepsilon_n^{-1})}}, \varepsilon_n < 1 \right\}, (\varepsilon_n \to 0, n \to \infty)$, viewed as a subset of $C[[0,1]]$, is $P_{\widetilde{\pi}}$-a.s. relative compact (in the uniform topology), and the set of its limit points is exactly K:*

$$K = \{x \in C \cap AC, x(0) = 0, \int\limits_0^1 (\frac{d}{dt}x(t))^2 dt \leq 1\},$$

where AC is the class of absolutely continuous functions on [0,1].

It is worth noticing that the above theorem can provide to us bounds for the converge of estimator of stationary probability and of course of asymptotic availability.

References

1. Aalen O, Borgan O, Gjessing H (2008) Survival and event history analysis. Springer, New York
2. Akritas MG, Roussas GG (1980) Asymptotic inference in continuous time semi-Markov processes. Scand J Stat 7:73–79
3. Anisimov VV (2008) Switching processes in queueing models. ISTE–J. Wiley, Applied Stochastic Methods Series, London
4. Asmussen S (1987) Applied probability and queues. Wiley, Chichester
5. Aven T, Jensen U (1999) Stochastic models in reliability. Springer, New York
6. Barbu V, Limnios N (2008) Semi-Markov chains and hidden semi-Markov models. Toward applications. Their use in reliability and DNA analysis. Lecture notes in statistics, vol 191. Springer, New York

7. Barbu V, Limnios N (2010) Some algebraic methods in semi-Markov chains. Contemp Math AMS 516:19–35
8. Barlow RE, Prochan F (1975) Statistical theory of reliability and life testing: probability models. Holt, Rinehart and Winston, New York
9. Chiquet J, Limnios N (2008) A method to compute the reliability of a piecewise deterministic Markov process. Stat Probab Lett 78:1397–1403
10. Chryssaphinou O, Karaliopoulou M, Limnios N (2008) On discrete time semi-Markov chains and applications in words occurrences. Commun Stat Theory Methods 37:1306–1322
11. Çinlar E (1969) Markov renewal theory. Adv Appl Probab 1:123–187
12. Çinlar E (1975) Introduction to stochastic processes. Prentice-Hall, Englewood Cliffs
13. Cox DR (1955) The analysis of non-Markovian stochastic processes by the inclusion of supplementary variables. Proc Camb Philos Soc 51:433–441
14. Csenki A (1995) Dependability for systems with a partitioned state space. Lecture notes in statistics, vol 90. Springer, Berlin
15. Devooght J (1997) Dynamic reliability. Adv Nucl Sci Technol 25:215–278
16. Esary JD, Marshal AW, Proschan F (1973) Shock models and wear processes. Ann Probab 1:627–649
17. Feller W (1964) On semi-Markov processes. Proc Natl Acad Sci USA 51(2):653–659
18. Gertsbakh I (2000) Reliability theory—with applications to preventive maintenance. Springer, Berlin
19. Gikhman II, Skorokhod AV (1974) Theory of stochastic processes, vol 2. Springer, Berlin
20. Gill RD (1980) Nonparametric estimation based on censored observations of a Markov renewal process. Z Wahrsch verw Gebiete 53:97–116
21. Girardin V, Limnios N (2006) Entropy for semi-Markov processes with Borel state spaces: asymptotic equirepartition properties and invariance principles. Bernoulli 12(3):515–533
22. Greenwood PE, Wefelmeyer W (1996) Empirical estimators for semi-Markov processes. Math Methods Stat 5(3):299–315
23. Grigorescu S, Oprişan G (1976) Limit theorems for J-X processes with a general state space. Z Wahrsch verw Gebiete 35:65–73
24. Gyllenberg M, Silvestrov DS (2008) Quasi-stationary phenomena in nonlinearly perturbed stochastic systems. de Gruyter, Berlin
25. Harlamov BP (2008) Continuous semi-Markov processes. ISTE–J. Applied Stochastic Methods Series, Wiley, London
26. Huzurbazar AV (2004) Flowgraph models for multistate time-to-event data. Wiley, New York
27. Iosifescu M, Limnios N, Oprişan G (2010) Introduction to stochastic models. Iste, Wiley, London
28. Janssen J (ed) (1986) Semi-Markov models. Theory and applications. Plenum Press, New York
29. Janssen J, Limnios N (eds) (1999) Semi-Markov models and applications. Kluwer, Dordrecht
30. Janssen J, Manca R (2006) Applied semi-Markov processes. Springer, New York
31. Keilson J (1979) Markov chains models—rarity and exponentiality. Springer, New York
32. Korolyuk VS, Limnios N (2004) Average and diffusion approximation for evolutionary systems in an asymptotic split phase space. Ann Appl Probab 14(1):489–516
33. Korolyuk VS, Limnios N (2005) Stochastic systems in merging phase space. World Scientific, Singapore
34. Korolyuk VS, Turbin AF (1993) Mathematical foundations of the state lumping of large systems. Kluwer, Dordrecht
35. Korolyuk VS, Swishchuk A (1995) Semi-Markov random evolution. Kluwer, Dordrecht
36. Kovalenko IN, Kuznetsov NYu, Pegg PA (1997) Mathematical theory of reliability of time dependent systems with practical applications. Wiley, Chichester
37. Lagakos SW, Sommer CJ, Zelen M (1978) Semi-Markov models for partially censored data. Biometrika 65(2):311–317
38. Limnios N (2004) A functional central limit theorem for the empirical estimator of a semi-Markov kernel. J Nonparametr Stat 16(1–2):13–18

39. Limnios N (2006) Estimation of the stationary distribution of semi-Markov processes with Borel state space. Stat Probab Lett 76:1536–1542
40. Limnios N (2010) Semi-Markov processes and hidden models, EORMS Encyclopedia of operation research and management science. Wiley, New York
41. Limnios N (2011) Reliability measures of semi-Markov systems with general state space. Methodology and Computing in Applied Probability, doi: 10.1007/s11009-011-9211-5
42. Limnios N, Nikulin M (eds) (2000) Recent advances in reliability theory: methodology, practice and inference. Birkhäuser, Boston
43. Limnios N, Oprişan G (2001) Semi-Markov processes and reliability. Birkhäuser, Boston
44. Limnios N, Ouhbi B (2003) Empirical estimators of reliability and related functions for semi-Markov systems. In: Lindqvist B, Doksum K (eds) Mathematical and statistical methods in reliability. World Scientific, Singapore
45. Limnios N, Ouhbi B (2006) Nonparametric estimation of some important indicators in reliability for semi-Markov processes. Stat Methodol 3(4):341–350
46. Lisnianski A, Levitin G (2003) Multi-state system reliability. World Scientific, Singapore
47. Moore EH, Pyke R (1968) Estimation of the transition distributions of a Markov renewal process. Ann Inst Stat Math 20:411–424
48. Osaki S (1985) Stochastic system reliability modeling. World Scientific, Singapore
49. Ouhbi B, Limnios N (1999) Nonparametric estimation for semi-Markov processes based on its hazard rate functions. Stat Inference Stoch Process 2(2):151–173
50. Ouhbi B, Limnios N (2003) Nonparametric reliability estimation of semi-Markov processes. J Stat Plann Inference 109(1/2):155–165
51. Ouhbi B, Limnios N (2002) The rate of occurrence of failures for semi-Markov processes and estimation. Stat Probab Lett 59(3):245–255
52. Pérez-Ocón R, Torres-Castro I (2002) A reliability semi-Markov model involving geometric processes. Appl Stoch Models Bus Ind 18:157–70
53. Prabhu NU (1980) Stochastic storage processes. Springer, Berlin
54. Pyke R (1961) Markov renewal processes: definitions and preliminary properties. Ann Math Stat 32:1231–1242
55. Pyke R (1961) Markov renewal processes with finitely many states. Ann Math Stat 32:1243–1259
56. Pyke R, Schaufele R (1964) Limit theorems for Markov renewal processes. Ann Math Stat 35:1746–1764
57. Pyke R, Schaufele R (1967) The existence and uniqueness of stationary measures for Markov renewal processes. Ann Stat 37:1439–1462
58. Revuz D (1975) Markov chains. North-Holland, Amsterdam
59. Ross SM (1970) Applied probability models with optimization applications. Holden-Day, San Francisco
60. Shurenkov VM (1984) On the theory of Markov renewal. Theory Probab Appl 29:247–265
61. Silvestrov DS (1996) Recurrence relations for generalized hitting times for semi-Markov processes. Ann Appl Probab 6(2):617–649
62. Trevezas S, Limnios N (2009) Maximum likelihood estimation for general hidden semi-Markov processes with backward recurrence time dependence. In: Notes of scientific seminars of the St. Petersburg Department of the Steklov Mathematical Institute, Russian Academy of Sciences, vol 363, pp 105–125, and J Math Sci, special issue in honor of I. Ibragimov
63. Wilson A, Limnios N, Keller-McNulty S, Armijo Y (eds) (2005) Modern statistical and mathematical methods in reliability. Series on quality, reliability, engineering statistics. World Scientific, Singapore

Chapter 7
Hazard Regression Analysis

7.1 Introduction: A Review of Regression Models for Lifetime Data

The physical description of the deterioration process of a system may require the consideration of several (exogenous and endogenous) factors that are commonly referred to as *covariates* or explanatory variables. The inclusion of this kind of information in the deterioration model may be attempted in several ways, for which reason we propose to study different regression models for lifetime data. There is a vast literature on semi-parametric models that addresses the relationship between lifetime and covariates (see, for example, the books by Andersen et al. [5], Klein and Moeschberger [32], Kleinbaum and Klein [33], Martinussen and Scheike [43] or Therneau and Grambsch [57]).

In this chapter, a common strategy is adopted: dependence on auxiliary information is managed through the hazard function as in Wang [67]. In other words, the instantaneous risk of failure of a particular device will be formulated in terms of the characteristics describing the item. Thus, we define the conditional hazard function with the following definition.

Definition 7.1 (*Conditional Hazard Function*) Let T be a random variable that denotes the lifetime of a system or device. Let $\mathbf{X} = (X_1, X_2, \ldots, X_p)^\top$ be a vector of p covariates with density function $\phi_\mathbf{X}$, the conditional hazard function of T given $\mathbf{X}$ is defined as

$$\lambda(t; \mathbf{x}) = \lim_{\Delta \to 0} \frac{P\{t < T \le t + \Delta \mid T > t, \mathbf{X} = \mathbf{x}\}}{\Delta}. \tag{7.1}$$

M. L. Gámiz et al., *Applied Nonparametric Statistics in Reliability*,
Springer Series in Reliability Engineering, DOI: 10.1007/978-0-85729-118-9_7,
© Springer-Verlag London Limited 2011

For a given $t > 0$ and given $\mathbf{x}$, the hazard function can be written as the ratio of the conditional density function $f(t; \mathbf{x})$ to the conditional survival function $S(t; \mathbf{x}) = 1 - F(t; \mathbf{x})$, that is

$$\lambda(t; \mathbf{x}) = \frac{f(t; \mathbf{x})}{S(t; \mathbf{x})},$$

for $S(t; \mathbf{x}) > 0$.

All the methods included in this chapter account for censoring, which implies a drastic limitation in the traditional models applied in standard statistical problems. The presence of right-censoring is probably the most common feature of datasets in reliability and survival studies and implies the termination of the observation of system-life due to causes other than the natural failure to which the system is subject. Let us formalize the conditions under which the methods presented below are established.

Assumption 7.1 (*Random Right-Censoring Model (RCM)*) *Consider that we have a sample consisting of n observations of type $\{(Y_1, \delta_1, \mathbf{X}_1), (Y_2, \delta_2, \mathbf{X}_2), ..., (Y_n, \delta_n, \mathbf{X}_n)\}$, with $Y_i = \min\{T_i, C_i\}$ for each $i = 1, 2, ..., n$, where*

- *$T_1, T_2, ..., T_n$ are independent realizations of a lifetime random variable T;*
- *$C_1, C_2, ..., C_n$ are independent realizations of a censoring random variable C;*
- *$\delta_1, \delta_2, ..., \delta_n$ are observations of the random variable $\delta = I\{Y = T\}$ where $Y = \min\{T, C\}$. This variable is usually referred to as the censoring indicator.*
- *$\mathbf{X}_1, \mathbf{X}_2, ..., \mathbf{X}_n$ are observations of a random vector of covariates $\mathbf{X}$;*
- *For a specific covariate vector $\mathbf{x}$, we have that T and C are conditionally independent given $\mathbf{X} = \mathbf{x}$.*

Random right-censoring is a particular case of what is called independent censoring assumption, which means that, conditional on covariates, the censored items are representative of those still at risk at the same time. In other words, failure rates of individuals at risk are the same as if there was no censoring, and hence, conditional on the covariates, items are not being censored because they have a higher or lower risk of failure. For a more detailed discussion, see Kalbfleisch and Prentice [31], pg. 12–13.

Under RCM, if we denote by $F(\cdot; \mathbf{x})$, $G(\cdot; \mathbf{x})$ and $H(\cdot; \mathbf{x})$ the conditional distribution functions of T, C and Y, respectively, given $\mathbf{X} = \mathbf{x}$, we have $H(\cdot; \mathbf{x}) = 1 - (1 - F(\cdot; \mathbf{x}))(1 - G(\cdot; \mathbf{x}))$.

Finally, we also assume that *noninformative censoring* is implicit in our model. With this, we mean that the censoring distribution function does not contain any information about the unknown lifetime distribution. Under a parametric approach, this means that the censoring distribution does not involve any of the unknown parameters in the model. In consequence, the part of the likelihood function that involves probabilities computed in terms of the distribution function $G(\cdot; \mathbf{x})$ is ignored in the estimation procedure.

Possibly the best known of all semi-parametric models for the conditional hazard function is the extensively used *Cox proportional hazards model*, which assumes proportionality of the hazard functions of two items defined by different sets of covariates. This assumption may, in many cases, be very restrictive, especially for long run modeling. Therefore, several alternatives have recently been proposed for modeling survival data where the proportional hazards assumption does not hold. The most popular in the field of reliability are the *Aalen additive model* and the *accelerated failure time model*. The less informative situation arises when no structure is considered in the function (7.1), which leads us to the nonparametric hazard models. Given a vector of covariates, the nonparametric estimation of the hazard rate may be tackled in several ways. The most usual approach is by smoothing (given a vector of covariates) the Nelson–Aalen estimator in two directions, first in the time argument and subsequently in the covariates. Other approaches have developed an estimator of the conditional hazard rate as the ratio of nonparametric estimators of a conditional density and a survival function.

7.1.1 The Cox Proportional Hazards Model

In biomedical research, the knowledge of factors that determine the prognosis of patients is of major clinical importance. In most cases, the response variable represents, in a sense, a survival time (i.e. the time that elapses before the occurrence of a particular event of interest), and therefore a regression model is formulated in order to determine the relationship between time and a set of explanatory variables. The Cox regression model is the model used by far the most in biostatistics applications and, generally, in survival and reliability studies. In our context of reliability, the survival time is interpreted as the time before failure occurs in a given device (system or component), and the aim is to evaluate this time in terms of the particular characteristics of the device.

Let T denote the random variable time-to-failure and $\mathbf{X} = (X_1, \ldots, X_p)^\top$ be a p-dimensional vector of covariates or explanatory variables that describe a particular device or system in terms of exogenous (such as temperature and pressure or, in general, the conditions describing the external environment in which the device works) and/or endogenous characteristics (such as size in the sense of physical dimensions, type of material the device is made of, etc.).

The basic model assumes that the hazard rate function of the failure time of a system with covariate vector given by $\mathbf{X}$ may be expressed as

$$\lambda(t; \mathbf{X}) = \lambda_0(t) \Psi(\boldsymbol{\beta}^\top \mathbf{X}) \tag{7.2}$$

where $\lambda_0(t)$ is an unspecified hazard function; $\boldsymbol{\beta}^\top = (\beta_1, \ldots, \beta_p)$ is a p-dimensional parameter vector; and $\Psi(\cdot)$ is a known function. The model does not assume any particular parametric form for $\lambda_0(t)$, referred to as the *baseline hazard function*. This function represents the hazard of an item with covariate vector equal to 0 (provided that $\Psi(0) = 1$), referred to as a baseline item. In this model, no assumption is made about the distribution of the failure time of the population of baseline items. So, this is a semi-parametric model in the sense that a parametric form is assumed for the covariate effect. In fact, a common model for $\Psi(\boldsymbol{\beta}^\top \mathbf{X})$ is

$$\Psi(\boldsymbol{\beta}^\top \mathbf{X}) = \exp(\boldsymbol{\beta}^\top \mathbf{X}) = \exp\left(\sum_{j=1}^{p} \beta_j X_j\right). \tag{7.3}$$

This regression model is also called a *proportional hazards model* and was introduced by Cox [13]. Basically, the model assumes that a proportional relationship exists between the hazard functions of failure times corresponding to different items. In other words, if we consider two devices, defined respectively by the covariate vectors $\mathbf{X}^1$ and $\mathbf{X}^2$, the ratio of the corresponding hazard functions is

$$\frac{\lambda(t; \mathbf{X}^1)}{\lambda(t; \mathbf{X}^2)} = \frac{\lambda_0(t) \exp\left(\sum_{j=1}^{p} \beta_j X_j^1\right)}{\lambda_0(t) \exp\left(\sum_{j=1}^{p} \beta_j X_j^2\right)} = \exp\left[\sum_{j=1}^{p} \beta_j \left(X_j^1 - X_j^2\right)\right] \tag{7.4}$$

which is constant with time. The hazards ratio in (7.4) is referred to, in biostatistics contexts, as the *relative risk* of an individual with risk factor $\mathbf{X}^1$ experiencing the event of interest (death or relapse, for instance) when compared to an individual with risk factor $\mathbf{X}^2$. We will also adopt, in our context, the denomination of relative risk for the quantity given in (7.4).

The prime interest is to make inferences about the parameter vector $\boldsymbol{\beta}$, which represents the *log-relative risk*, and the baseline hazard function $\lambda_0(t)$ or the cumulative baseline hazard function, that is, $\Lambda_0(t) = \int_0^t \lambda_0(u)\, du$.

Assume that we have n independent observations of the form $(Y_i, \delta_i, \mathbf{X}_i)$, $i = 1, 2, \ldots, n$, under the RCM specifications. That is, Y_i are right-censored lifetimes, which we assume are ordered; δ_i is the censoring indicator, which tells whether an observation is censored or not ($\delta_i = 1$ if the failure has occurred at Y_i and $\delta_i = 0$ if the lifetime is right-censored); and $\mathbf{X}_i$ is a vector of explanatory variables.

Estimation of the parameter $\boldsymbol{\beta}$ has traditionally been based on a *partial* or *conditional likelihood* formulation, where the baseline hazard is understood as a nuisance parameter, which is not, in general, estimated, since the aim is to evaluate the effect that each risk factor has over the risk of failure.

Let us define the *at-risk process* as $D(t) = I[Y \geq t]$, as given in Martinussen and Scheike [43]. The partial likelihood is obtained as the product, extended to all subjects in the sample, of the conditional probability that a subject with covariates $\mathbf{X}_i$ fails at time Y_i, given that one of the subjects at risk at Y_i fails at this time, that is

P [subject i fails at Y_i| one failure at Y_i]

$$
\begin{aligned}
&= \frac{P[\text{subject i fails at} Y_i | \text{at} - \text{risk at } Y_i]}{P[\text{one failure at } Y_i | \text{at-risk at } Y_i]} \\[2mm]
&= \frac{\lambda(Y_i | \mathbf{X}_i)}{\sum_{j=1}^{n} D_j(Y_i)\lambda(Y_i | \mathbf{X}_j)} \\[2mm]
&= \frac{\lambda_0(Y_i) \exp[\boldsymbol{\beta}^\top \mathbf{X}_i]}{\sum_{j=1}^{n} D_j(Y_i)\lambda_0(Y_i) \exp[\boldsymbol{\beta}^\top \mathbf{X}_j]} \\[2mm]
&= \frac{\exp[\boldsymbol{\beta}^\top \mathbf{X}_i]}{\sum_{j=1}^{n} D_j(Y_i) \exp[\boldsymbol{\beta}^\top \mathbf{X}_j]},
\end{aligned}
\tag{7.5}
$$

where $D_j(t) = I[Y_j \geq t]$. The partial likelihood is then constructed as

$$
PL(\boldsymbol{\beta}) = \prod_{i=1}^{n} \left[\frac{\exp[\boldsymbol{\beta}^\top \mathbf{X}_i]}{\sum_{j=1}^{n} D_j(Y_i) \exp[\boldsymbol{\beta}^\top \mathbf{X}_j]} \right]^{\delta_i}
\tag{7.6}
$$

The flexibility of the model (7.2) lies in the nonparametric term, $\lambda_0(t)$, the baseline hazard function. The estimation procedure of this term is based on a profile likelihood, which is constructed by fixing a value of the regression parameter $\boldsymbol{\beta}$ which is then maximized in λ_0. Thus, we obtain the profile maximum likelihood estimator of $\lambda_0(Y_i)$ as

$$
\hat{\lambda}_{0i} = \frac{\delta_i}{\sum_{j=1}^{n} D_j(Y_i) \exp[\boldsymbol{\beta}^\top \mathbf{X}_j]}.
\tag{7.7}
$$

This estimator of the baseline hazard rate leads us to the following estimator of the corresponding cumulative hazard rate known as the Breslow estimator ([7]) and given by

$$
\hat{\Lambda}_0(t) = \sum_{i:Y_i \leq t} \frac{\delta_i}{\sum_{j=1}^{n} D_j(Y_i) \exp[\boldsymbol{\beta}^\top \mathbf{X}_j]}.
\tag{7.8}
$$

A large number of specialized books exist in the recent literature, which include an extensive and comprehensive treatment of the Cox proportional hazards model. We particularly recommend the text of Klein and Moeschberger [32] where all the methods presented are properly illustrated by means of numerous practical examples in the context of biomedical applications.

Although traditionally biostatistics is the natural application area for proportional hazards modeling, this method has progressively achieved more and more prominence in the field of reliability engineering, and as a consequence, the number of papers that illustrate the use of this model under different features has increased in the recent literature on reliability modeling. As an example, Carrion et al. [8] present a very simple application of the proportional hazards model in the study of breakdown of pipes of a water supply network system.

The relevance of this model strongly relies in that, parallel to the development of important theoretical results in the recent years, there is the implementation of the procedures in the available software. The majority of the existing statistical packages account for functions that make it easy to fit and check the validity of Cox models in real applications. The free software environment R is nowadays possibly the leading software in this respect, and, in particular, the package *survival* [23, 58] that provides several functions and datasets for survival analysis. In Sect. 7.2.1 we briefly present an illustrative example of the use of some procedures included in this package to discuss a particular Cox model fitting.

7.1.2 The Aalen Additive Model

When the Cox model does not fit correctly to a dataset, one alternative is given by the additive model proposed by Aalen [1–3] where

$$\lambda(t; X_1(t), \ldots, X_p(t)) = \beta_0(t) + \beta_1(t)X_1(t) + \cdots + \beta_p(t)X_p(t) \qquad (7.9)$$

where $\mathbf{X}(t) = (X_1(t), \ldots, X_p(t))^{\top}$ denotes a vector of possibly time-dependent covariates. So, the hazard function is considered a linear function of the covariates with coefficients that are allowed to vary with time. In this approach, $\beta_0(t)$ is a baseline hazard function and $\beta_j(t)$ is the regression function that describes the effect of covariate X_j over the risk of failure of an item at a given time t. Assume that we have n independent observations of the form $(Y_i, \delta_i, \mathbf{X}_i)$, $i = 1, 2, \ldots, n$, where $\mathbf{X}_i = \mathbf{X}_i(Y_i)$. Define the at-risk process as in the previous section, that is $D(t) = I[Y \geq t]$. Then, associated with each observed item, we have $D_i(t) = I[Y_i \geq t]$. Therefore, with the additive hazard regression model (7.9), the conditional hazard rate may be expressed in matricial notation as

$$\lambda(t) = \mathbf{D}(t)\widetilde{\mathbf{X}}(t)\boldsymbol{\beta}(t) \qquad (7.10)$$

where we call $\lambda(t) = (\lambda_1(t), \lambda_2(t), \ldots, \lambda_n(t))^{\top}$, $\lambda_i(t)$ being the intensity at which failure occurs for the ith item, for $i = 1, 2, \ldots, n$; $\mathbf{D}(t)$ is an n dimensional diagonal matrix whose elements are the $D_i(t)$, that is, $\mathbf{D}(t) = \text{diag}(D_1(t), D_2(t), \ldots, D_n(t))$; and, $\widetilde{\mathbf{X}}(t)$ is a $n \times (p + 1)$ matrix whose ith row is $\widetilde{\mathbf{X}}_i(t)^{\top} = (1, \mathbf{X}_i(t)^{\top}) = (1, X_{i1}(t), X_{i2}(t), \ldots, X_{ip}(t))$ a $p + 1$ dimensional vector with the covariates defining the ith sample element. The first component of this vector is set to 1 to allow for a baseline hazard rate, that is, the baseline effect is considered in the vector of unknowns $\boldsymbol{\beta}(t) = (\beta_0(t), \beta_1(t), \ldots, \beta_p(t))^{\top}$. This specification allows us to derive the full estimation procedure not only for the regression effects but also for parameters describing the baseline hazard.

The estimation procedures of this model are based on the least squares method. Since direct estimation of $\boldsymbol{\beta}(t)$ can be difficult, it is more feasible to estimate the cumulative regression function denoted by $\mathbf{B}(t)$ whose components are defined as

$$B_j(t) = \int_0^t \beta_j(u)\,du, \tag{7.11}$$

for $j = 0, 1, \ldots, p$.

Let us consider that the observed times at which events (failure or censoring) occur are ordered so that $Y_1 < Y_2 < \cdots < Y_n$. Then, reasoning in a similar way to that used in the usual linear models, the least squares estimator of $\mathbf{B}(t)$ is given by the generalized inverse of $\mathbf{D}(t)\widetilde{\mathbf{X}}(t)$ and can be expressed

$$\widehat{\mathbf{B}}(t) = \sum_{i:Y_i \leq t} \left[\left(\mathbf{D}(Y_i)\widetilde{\mathbf{X}}(Y_i)\right)^{\top}\mathbf{D}(Y_i)\widetilde{\mathbf{X}}(Y_i) \right]^{-1} \left(\mathbf{D}(Y_i)\widetilde{\mathbf{X}}(Y_i)\right)^{\top}\mathbf{e}_i \tag{7.12}$$

where $\mathbf{e}_i$ denotes an n-dimensional vector whose ith element is δ_i, i.e. the censoring indicator and the rest of the elements are all equal to 0.

The equation above may be simplified so that

$$\widehat{\mathbf{B}}(t) = \sum_{i:Y_i \leq t} \left[\widetilde{\mathbf{X}}(Y_i)^{\top}\mathbf{D}(Y_i)\widetilde{\mathbf{X}}(Y_i) \right]^{-1} \widetilde{\mathbf{X}}(Y_i)^{\top}\mathbf{e}_i. \tag{7.13}$$

This estimator is only defined for $t \leq \tau$, where τ is the largest value of Y_i for which matrix $\mathbf{M}(t) = \widetilde{\mathbf{X}}(t)^{\top}\mathbf{D}(t)\widetilde{\mathbf{X}}(t)$ is nonsingular.

Crude estimates of the regression coefficient $\beta_j(t)$ can be obtained by examining the slope of the estimated $B_j(t)$. It is possible to obtain more flexible estimates of the regression functions by using kernel smoothing techniques.

For a detailed development of the model presented in this section, the reader is referred to the book of Martinussen and Scheike [43], where numerous practical examples in the context of biomedical applications are discussed. Moreover, these authors have recently contributed to the public domain R by incorporating the package *timereg* [53] which, among other things, includes the function *aalen* for fitting the additive hazards model of Aalen.

7.1.3 The Accelerated Failure Time Model

The accelerated life model relates the logarithm of the lifetime T linearly to the vector of covariates $\mathbf{X}$. Specifically it can be written as

$$\ln T = \psi(\mathbf{X}) + \varepsilon, \tag{7.14}$$

where ε is a random error term and ψ is an unknown function. This model forms part of the hazard regression family of models. In fact, we have the following sequence of equalities

$$P\{T > t\} = P\{\ln T > \ln t\} = P\{\varepsilon > \ln(t\exp(-\psi(\mathbf{X})))\}$$
$$= P\{T_0 > t\exp(-\psi(\mathbf{X}))\}, \tag{7.15}$$

where it is useful to introduce the non-negative random variable $T_0 = \exp(\varepsilon)$. Looking at the relation between T and T_0, it is true that

$$\Lambda(t) = \Lambda_0(t\exp(-\psi(\mathbf{X})),$$

where Λ_0 is the cumulative hazard function corresponding to T_0. This last equation can be written in terms of the corresponding hazard functions, then, if we call $\Psi(\mathbf{X}) = \exp(-\psi(\mathbf{X}))$, we have

$$\lambda(t) = \lambda_0(t\Psi(\mathbf{X}))\Psi(\mathbf{X}),$$

which establishes that the accelerated failure model does not hold proportionality between hazard rates (except when we have a Weibull regression model, which is when $\lambda_0(t) = \alpha\gamma t^{\gamma-1}$, for suitable α, γ), thus providing an interesting alternative to the Cox proportional hazards model. In fact in some cases, it constitutes a more appealing model option than the proportional hazards model, due to its direct physical interpretation.

In this context, accelerated life tests [45, 47, 48], are increasingly being used in manufacturing industries. Accelerated life testing is a method that exposes items to higher stress levels than they would receive during their normal use. The main objective is to induce early failure and the motivation to do this lies in the fact that, when the mean lifetime of a device is measured in decades (for example) under normal conditions of use, it would be necessary to wait many years to establish the degree of reliability of such a device. The advances in current technology are nowadays so rapid that a particular device may become obsolete before its reliability properties can be determined by testing in normal use conditions.

The conditions of use of a system are typically expressed in terms of so-called stress factors, such as temperature, voltage, pressure and humidity. The accelerated life test is run at high levels of these factors (significantly higher levels than those in normal conditions) in order to force reduced times to failure of the system. The goal consists of inferring the reliability properties of the system at a normal level of stress by viewing its behavior at an accelerated stress level. To do this, it is essential to use a model that captures the direct relationship between lifetime and stress, which makes the accelerated failure time model the best option.

In the semi-parametric approach, the model can assume a physically interpretable relationship between lifetime and stress levels, with no assumption being made about the distribution of the lifetime [16]. Consider a p-dimensional stress vector $\mathbf{X} = (X_1, X_2, \ldots, X_p)^\top$. Let T_0 denote the random time-to-failure at a normal use stress level, with S_0 as the corresponding survival function. Let T, on the other hand, denote the lifetime at the accelerated stress level $\mathbf{X}$. According to (7.15) we have

$$S(t) = S_0(t\exp(\psi(\mathbf{X}))) = S_0(t\Psi(\mathbf{X})). \tag{7.16}$$

The main purpose is to estimate S_0 from observations of the lifetime at accelerated stress levels. The function $\Psi(\mathbf{X})$ is referred to as the *acceleration factor*.

Most accelerated failure time models assume a linear function of a constant stress covariate, leading to the following semi-parametric model in the logarithmic scale of lifetime

$$\ln T = \boldsymbol{\beta}^{\top}\mathbf{X} + \varepsilon, \tag{7.17}$$

where ε is usually assumed to have a distribution with location parameter 0 and scale parameter σ. Special cases that are very often considered are the exponential, Weibull or lognormal distributions. With this focus, the log-lifetime is considered to have a distribution with location parameter $\mu(\mathbf{X}) = \boldsymbol{\beta}^{\top}\mathbf{X}$ and scale parameter σ, where the unknown parameters are estimated from the accelerated test data. Thus, the location parameter of the log-time, μ, is a simple linear function of the stress variable that could be previously transformed with regard to certain physical arguments as considered in the formulation of, among others that are widely used in practice, the Arrhenius model, inverse power model and exponential model. Using this approach and in the particular case of a single covariate, one could express the model as

$$\tilde{S}(u; x, \beta_0, \beta_1, \sigma) = \tilde{S}_0\left(\frac{u - \beta_0 + \beta_1 x}{\sigma}\right), \tag{7.18}$$

where $\tilde{S}$ and $\tilde{S}_0(\cdot/\sigma)$ are the respective survival functions of $\ln T$ and $\varepsilon = \ln T_0$. Typically, the main purpose is to estimate a specified percentile of the distribution of the lifetime in conditions of use, say x_0, which can be denoted as $t_\pi(x_0)$, for $0 < \pi < 1$. For example, the interest is usually centered at the median lifetime. We can express

$$t_\pi(x_0) = \beta_0 + \beta_1 x_0 + \tilde{u}_\pi \sigma, \tag{7.19}$$

$\tilde{u}_\pi$ being the corresponding percentile in the distribution given by $\tilde{S}_0$. The inference problem is then reduced to obtain suitable estimations $\hat{\beta}_0$, $\hat{\beta}_1$ and $\hat{\sigma}$. These models are studied in detail in Nelson [45] and Meeker and Escobar [44].

From this point of view, censored quantile regression approaches are now appealed. In particular, we will concentrate on the median regression model. In fact, although most of the works on this subject express the accelerated failure time model in terms of the mean and establish that the mean of the logarithm of the survival time is linearly related to the covariates, the bias caused by censoring suggests a more robust procedure, and therefore the median offers a suitable alternative. The censored median regression model has received much attention recently, see for instance, the works by Ying et al. [66], Yang [65], Honoré, Khan and Powell [28], Cho and Hong [11], Zhao and Chen [68], and Wang and Wang [61]. For a comprehensive presentation of the theory of quantile regression with noncensored data, the reader is referred to Koenker [34].

7.1.3.1 Censored Median Regression

The semi-parametric censored median regression model can be formulated as follows. Let T denote, as above, the time-to-failure or a convenient transformation of it. Under right-censoring, we observe $Y = \min\{T, C\}$. Consider a random sample $\{(Y_i, \delta_i, \mathbf{X}_i), i = 1, 2, \ldots, n\}$, under the conditions specified in the RCM. Given a vector of covariates $\mathbf{X}$, let us consider the semi-parametric model in terms of the median of the conditional log failure time $\widetilde{T}$, that is

$$\text{median}(\widetilde{T}|\mathbf{X}) = m(\mathbf{X}) = \beta_0 + \beta_1 X_1 + \cdots + \beta_p X_p = \boldsymbol{\beta}^\top \mathbf{X}. \tag{7.20}$$

In this formulation, the intercept term is introduced to allow the assumption that the random error in (7.14), i.e. ε has a distribution with median equal to zero.

In the case of noncensoring, the median is defined as the minimizer of the least absolute deviation function, and the β_i's can then be estimated by minimizing the sum of the absolute values of the errors, that is

$$\widehat{\beta} = \arg\min_\mathbf{b} \left\{ \frac{1}{n} \sum_{i=1}^n |\widetilde{T}_i - \mathbf{b}^\top \mathbf{X}_i| \right\}. \tag{7.21}$$

This is a particular case into the wider context of quantile regression. For a known value $\pi \in (0,1)$, in case of uncensored observations, Koenker and Bassett [36] proposed a quantile regression estimator as

$$\widehat{\beta}_\pi = \arg\min_\mathbf{b} \left\{ \frac{1}{n} \sum_{i=1}^n \rho_\pi(\widetilde{T}_i - \mathbf{b}^\top \mathbf{X}_i) \right\}, \tag{7.22}$$

with $\rho_\pi(u) = u(\pi - I[u<0])$. It can be shown, see Koenker [34] for further details, that a minimizer of the function in (7.22) is equivalent to a root of the following set of estimating equations

$$U_n(\mathbf{b}) = \sum_{i=1}^n \mathbf{X}_i \{ \pi - I[\widetilde{T}_i - \mathbf{b}^\top \mathbf{X}_i \leq 0] \} \approx 0. \tag{7.23}$$

where the symbol $\approx$ is used because $U_n(\mathbf{b})$ is a discontinuous function of $\mathbf{b}$. In the special case $\pi = 1/2$, function ρ is equivalently represented by the absolute value function, and we have the following estimating equations

$$U_n(\mathbf{b}) = \sum_{i=1}^n \mathbf{X}_i \left\{ \frac{1}{2} - I[\widetilde{T}_i - \mathbf{b}^\top \mathbf{X}_i \leq 0] \right\} \approx 0. \tag{7.24}$$

In the case of censoring, we observe Y_i instead of T_i, and therefore $\widetilde{Y}_i = \ln Y_i$ instead of $\widetilde{T}_i$, so the idea is to replace the unobservable quantity $I[\widetilde{T}_i - \mathbf{b}^\top \mathbf{X}_i \leq 0]$ by an estimation of its conditional expectation, given the data. Define the following

$$E_i(\mathbf{b}) = E\{I[\widetilde{T}_i - \mathbf{b}^\top \mathbf{X}_i \leq 0] \mid (\widetilde{Y}_i, \delta_i, X_i)\}.$$

To obtain this expectation, the idea of Efron [17] of redistribution of mass to the right may be adopted, see [61, 66] and Portnoy [51]. Then, one may note that in the above expectation, the contribution of each data point depends only on the sign of $\widetilde{T}_i - \mathbf{b}^\top \mathbf{X}_i$, so we can distinguish the following cases

- for uncensored observations, that is when $\delta_i = 1$, $\widetilde{Y}_i = \widetilde{T}_i$, then, given the data, the expectation reduces to $I[\widetilde{Y}_i - \mathbf{b}^\top \mathbf{X}_i \geq 0]$;
- when $\delta_i = 0$ and $\widetilde{Y}_i = \widetilde{C}_i > \mathbf{b}^\top \mathbf{X}_i$ where we use the notation $\widetilde{C}_i = \ln C_i$, the conditional expectation can be represented by $I[\widetilde{Y}_i - \mathbf{b}^\top \mathbf{X}_i \leq 0]$, since in this case we exactly know the sign of $\widetilde{T}_i - \mathbf{b}^\top \mathbf{X}_i$;
- when $\delta_i = 0$ and $\widetilde{C}_i < \mathbf{b}^\top \mathbf{X}_i$ the conditional expectation is obtained as

$$E_i(\mathbf{b}) = I\left[\widetilde{C}_i < \mathbf{b}^\top \mathbf{X}_i\right] \frac{P(\widetilde{T}_i < \mathbf{b}^\top \mathbf{X}_i | \mathbf{X}_i)}{P(\widetilde{T}_i > \widetilde{Y}_i | \mathbf{X}_i)} + I\left[\widetilde{Y}_\infty < \mathbf{b}^\top \mathbf{X}_i\right] \frac{P(\widetilde{T}_i > \mathbf{b}^\top \mathbf{X}_i | \mathbf{X}_i)}{P(\widetilde{T}_i > \widetilde{Y}_i | \mathbf{X}_i)},$$

where $\widetilde{Y}_\infty$ denotes a value large enough.

Thus, we can write the minimization problem for the censored median regression as

$$\widehat{\beta} = \arg\min_{\mathbf{b}} \left\{ \sum_{i=1}^n \omega_i |\widetilde{Y}_i - \mathbf{b}^\top \mathbf{X}_i| + \sum_{i=1}^n \omega_i^* |\widetilde{Y}_\infty - \mathbf{b}^\top \mathbf{X}_i| \right\}, \quad (7.25)$$

[34] where, for $i = 1, 2, \ldots, n$, the weights ω_i and ω_i^* are given by

$$\omega_i = \begin{cases} 1, & \delta_i = 1 \quad \text{or} \quad \widetilde{Y}_i > \mathbf{b}^\top \mathbf{X}_i \\ \dfrac{\widetilde{S}(\widetilde{Y}_i | \mathbf{X}_i) - 0.5}{\widetilde{S}(\widetilde{Y}_i | \mathbf{X}_i)}, & \delta_i = 0 \quad \text{and} \quad \widetilde{Y}_i < \mathbf{b}^\top \mathbf{X}_i \end{cases} \quad \text{and} \quad \omega_i^* = 1 - \omega_i,$$

$$(7.26)$$

where $\widetilde{S}(\cdot | \mathbf{X})$ the survival function of $\widetilde{T}$ given $\mathbf{X}$. This function is unknown and could be approximated from the data by, for instance, the Kaplan–Meier method.

Hence, to estimate the vector of the regression parameters β we consider an artificial dataset of size $2n$, obtained by expanding the original sample as follows $\{(Y_i, \delta_i, \mathbf{X}_i), i = 1, 2, \ldots, n\} \cup \{(Y_\infty, \delta_i, \mathbf{X}_i), i = 1, 2, \ldots, n\}$. The advantage of the above approach is that we can use existing software to compute estimations of the coefficients in β. For example, we can use the function rq included into the package *quantreg* [35] of R for fitting the weighted median regression model by using the expanded dataset and considering weigths $\{\omega_i, \omega_i^*\}$.

7.1.3.2 Local Quantile Regression

Finally, we are assuming the hypothesis of linearity; however, this assumption could not hold globally. Since we are interested in the median, we can relax this hypothesis and require linear dependence between $\widetilde{T}$ and $\mathbf{X}$ at this particular quantile. Wang and Wang [61] propose estimating locally the weights ω_i by using the local Kaplan–Meier estimators, that is

$$\widehat{\widehat{S}}_h(t, \mathbf{X}) = \prod_{i=1}^{n} \left\{ 1 - \frac{I[\widetilde{Y}_i \leq t] \delta_i B_{ni}(\mathbf{X})}{\sum_{j=1}^{n} I[\widetilde{Y}_j \geq \widetilde{Y}_i] B_{nj}(\mathbf{X})} \right\}, \tag{7.27}$$

where $\{B_{ni}(\mathbf{X})\}_{i=1}^{n}$ is a sequence of non-negative weights adding up to 1, see [25]. When $B_{ni}(\mathbf{X}) = 1/n$ for all $i = 1, 2, \ldots, n$ we obtain in (7.27) the classical Kaplan–Meier estimator of the survival function of $\widetilde{T}$. Although other possibilities may be taken into account, we consider a Nadaraya–Watson type estimator as in [61] and define

$$B_{ni}(\mathbf{X}) = \frac{k\left(\frac{\mathbf{X}-\mathbf{X}_i}{h}\right)}{\sum_{j=1}^{n} k\left(\frac{\mathbf{X}-\mathbf{X}_i}{h}\right)}, \tag{7.28}$$

with k a multivariate density kernel function and $h > 0$ the bandwidth parameter.

Example Uncensored Case

The estimation of the coefficients involved in the median regression problem can be tackled by using the function *rq* of the R package *quantreg*. To illustrate the procedure, we first consider a case where there is not censoring present in the dataset. Specifically, we use the data in Exercise 17.3 (page 461) of Meeker and Escobar [44], which refers to a life test on rolling contact fatigue of ceramic ball bearing. Four levels of stress are considered, and 10 specimens are tested at each. The lifetimes are given in $10^6 revolutions$, whereas the stress is measured in units of $10^6 psi$. Figure 7.1 gives plots not only for the median function, which is given by the solid line, but also for the quantiles of several orders (π), specifically we have considered $\pi = 0.05, 0.25, 0.4, 0.6, 0.75, 0.95$, which are represented by the dotted lines in the plot. Finally, for comparison purposes, we have also added to the graph the results of fitting a linear model in terms of the mean log lifetime (dashed line). The function *lm* of R has been used here.

Example Censored Case

Meeker and Escobar [44] analyze a low-cycle fatigue life data corresponding to 26 specimens of a nickel-base superalloy. Each observation consists of the number of cycles until failure or removal from the test and a level of pseudostress. There are four censored observations in the dataset (which is available from the Appendix C in [44]). To analyze these data, we first follow the guidelines presented in Koenker [35], where an R code is developed to estimate the model suggested in Portnoy [51] for quantile regression under random censoring. After

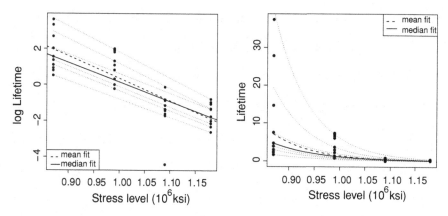

Fig. 7.1 Quantile regression: uncensored data

constructing the model by means of the function *crq*, we use the function *summary* to compute bootstrapped standard errors for the quantile regression estimates, see Koenker [35] for further details on this topic. The results are displayed in Fig. 7.2, where a 95% confidence region is represented by the shady region. A horizontal solid line is indicating a null effect of the covariate in the quantile process. The solid line with the circles represents the point estimate of the respective percentile regression fits.

As explained previously, the assumption of linearity may be too restrictive. So we have considered the local approach in this example following the steps detailed above. The results that we present concerns estimation for the regression model in terms of the three quartiles, that is for $\pi = 0.25, 0.5, 0.75$, which can be seen in

Fig. 7.2 Quantile regression: censored data

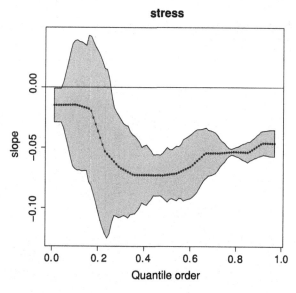

Fig. 7.3, respectively from left to right. Although, no differences can be appreciated for the case of the median and the third quartile, this is not the case for the first quartile where the fitted curves differ significatively.

7.1.4 More General Models

In a general setting, it can be assumed that the hazard function depends on the covariates as follows

$$\lambda\left(t,\mathbf{X}^1,\mathbf{X}^2\right) = \lambda_0\left(t,\mathbf{X}^1\right)\Psi\left(\mathbf{X}^2\right),\tag{7.29}$$

where the p-dimensional vector of covariates $\mathbf{X}$ has been partitioned in two subvectors $\mathbf{X}^1$ and $\mathbf{X}^2$ of dimensions p_1 and p_2, respectively.

In this framework, Dabrowska [15] was the first to extend the Cox proportional hazards model to allow baseline hazard functions dependent on covariates, by formulating the model

$$\lambda(t,\mathbf{X}^1,\mathbf{X}^2) = \lambda_0(t,\mathbf{X}^1)\exp\left(\boldsymbol{\beta}^\top\mathbf{X}^2\right)\tag{7.30}$$

This model can be considered as a generalization of the stratified Cox model (see Chap. 5 of [33], for a description). Let us suppose that the covariates included in vector $\mathbf{X}^1$ do not satisfy the proportional hazards assumption. For the sake of simplicity, think of a case where $p_1 = 1$ and then consider that X^1 is a single covariate with two levels. The model then consists of a Cox proportional hazards model with common regression coefficients for covariates included in $\mathbf{X}^2$, whereas two different baseline hazard functions are considered for each level of covariate X^1.

Fig. 7.3 Quantile regression: local estimate for censored data

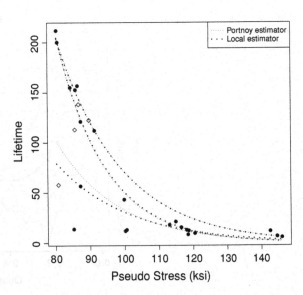

The models exposed in the previous sections can be viewed as particular cases of the Eq. (7.29). For example, in the absence of the vector of covariates $\mathbf{X}^1$, Eq. (7.29) reduces to the Cox proportional hazards model.

In a case where $\lambda_0(t, \overline{\mathbf{X}}^1) = \lambda_1 t \cdot \psi(\overline{\mathbf{X}}^1)$, for suitable λ_1 and ψ_1 we have that Eq. (7.29) reduces to the accelerated failure time model when $\mathbf{X}^1 = \mathbf{X}^2$ and $\varphi = \psi$. On the other hand, in the absence of the vector of covariates $\mathbf{X}^2$ it reduces to the so-called accelerated hazards model introduced by Chen and Wang [10]. This model is nonproportional and, depending on the shape of the baseline hazard function, can accommodate crossing hazard functions. A more general version of this model was formulated by Chen and Jewell [9]

$$\lambda(t, \mathbf{X}) = \lambda_0 \big(t \exp(\boldsymbol{\beta}_1^\top \mathbf{X}) \big) \exp(\boldsymbol{\beta}_2^\top \mathbf{X}). \tag{7.31}$$

In this case, the covariate effects are jointly determined by the two-dimensional parameter $\boldsymbol{\beta}^\top = (\boldsymbol{\beta}_1^\top, \boldsymbol{\beta}_2^\top)$. On the one hand, $\boldsymbol{\beta}_1$ can be interpreted as the acceleration or deceleration of the hazard progression caused by the magnitude of the covariate, while $\boldsymbol{\beta}_2$ identifies the relative risk after adjusting for the different hazard progressions at each level of the covariate. Of course, model (7.31) includes as subclasses the proportional hazards model, the accelerated failure time model and the accelerated hazards model. Moreover, it can itself be considered as a particular case of the model proposed by Dabrowska (Eq. 7.30), for $\mathbf{X}^1 = \mathbf{X}^2$. The authors check the identifiability of the model except when the underlying distribution is Weibull, as in this case, all the subclasses and the general model coincide.

Finally, some covariates can be considered to act additively over the hazard function, while others have a multiplicative effect. This leads us to the Cox-Aalen model which is expressed as

$$\lambda(t, \mathbf{X}^1, \mathbf{X}^2) = \Big(\boldsymbol{\beta}_1(t)^\top \mathbf{X}^1 \Big) \exp\Big(\boldsymbol{\beta}_2(t)^\top \mathbf{X}^2 \Big), \tag{7.32}$$

see for intance [43].

Remark 7.1 Variable Selection in the Cox Model   One of the accompanying issues with regression modeling is the variable selection and model building. Several classical variable selection and model building techniques like forward selection, backward selection and stepwise selection methods can be used in variable selection in the Cox proportional hazard model. In typical applications of these methods, a hypothesis test using a score statistic is conducted at each stage using a predetermined "significance level" or using a p-value. For example, for the data with $N_i(t)$ taking the value one if the item i, $i = 1, \ldots, n$ has been observed to fail before time t and taking the value zero otherwise and $D_i(t)$ is the at-risk process for the ith item, the Cox partial likelihood score vector is given by

$$u(\boldsymbol{\beta}) = \sum_{i=1}^{n} \int_0^\infty \big[\mathbf{X}_i - \overline{\mathbf{X}}(s, \boldsymbol{\beta}) \big] \, \mathrm{d}N_i(s)$$

where $dN_i(s) = N_i(s) - N_i(s-)$ and

$$\bar{\mathbf{X}}(s,\boldsymbol{\beta}) = \frac{\sum_{j=1}^n \mathbf{X}_j D_j(s) e^{\boldsymbol{\beta}'\mathbf{X}_j}}{\sum_{j=1}^n D_j(s) e^{\boldsymbol{\beta}'\mathbf{X}_j}}$$

Now, an asymptotic score test can be constructed using a statistic of the type $T = u(\boldsymbol{\beta})' I^{-1} u(\boldsymbol{\beta})$ where I is a version of the information matrix obtained via the second derivative of the log-Cox likelihood evaluated at the estimated parameter vector under each corresponding null hypothesis. The asymptotic critical points for this statistic can be obtained using a Chi-square distribution. Lawless [41] and references therein give some modifications to the score statistic and the resulting score tests. The popular PHREG procedure in SAS can be used to conduct a variable selection using any of the aforementioned classical methods. A rather new forward selection technique for the Cox model using pseudo explanatory variables has been proposed by Boos et al. [6]. Their proposal is a variant of a method called False Selection Rate method of Wu et al. [64], and their proposed method has been shown to have computational simplicity and high selection accuracy.

As in the linear regression, classical methods have several draw backs in selecting variables. One major hindrance is the lack of stability. That is, a slight perturbation of the data can lead to an entirely different model. In addition, model selection and parameter estimation are typically done separately in classical variable selection. To avoid these difficulties, the model selection using criteria that involves a penalty (regularization methods) have become a popular tool in modern selection problems. Penalized methods have two main orientations. One direction is to penalize for the number of terms alone; AIC (Akaike [4]), C_p (Mallows [42]), BIC (Schwarz [54]), RIC (Shi and Tsai [55]) etc. The other orientation is to use roughness penalties on the magnitude of the coefficients; LASSO (Tibshirani [59]), SCAD (Fan and Li [19]), Adaptive LASSO (Zou [69]) to mention a few popular techniques. In all penalty-based methods, the variable selection and estimation is done simultaneously.

An extension of regularization to the Cox model has been discussed by Fan and Li [20] using a nonconcave penalty. They first consider the penalized log likelihood

$$M(\boldsymbol{\beta}, \gamma) = \sum_{i=1}^n \{\delta_i[\ln \lambda_0(Y_i) + \boldsymbol{\beta}'\mathbf{X}_i] - \Lambda_0(Y_i)\exp\{\boldsymbol{\beta}'\mathbf{X}_i\}\} - n\sum_{j=1}^d p_\gamma(|\beta_j|)$$

for a d dimensional covariate vector $\boldsymbol{\beta}$ with a penalty factor γ. Here, p_γ is a suitable penalty function such as the SCAD penalty function [19]. Since the baseline hazard rate λ_0 is unknown, maximization of M with respect to $\boldsymbol{\beta}$ is not possible. Hence, Fan and Li [20] suggest a parameterization of the Λ_0 in a suitable manner. For example, if $t_j^0, j = 1, ..., N$ are the observed distinct lifetimes, one can let

$$\Lambda_0(t) = \sum_{j=1}^N \lambda_j I[t_j^0 \le t]$$

and maximize the resulting approximate log likelihood

$$\tilde{M}(\lambda_1, \ldots, \lambda_N, \gamma) = \sum_{j=1}^{N} \left[\ln \lambda_j + \boldsymbol{\beta}' \mathbf{X}_j^* \right] - \sum_{i=1}^{n} \sum_{j=1}^{N} \lambda_j I[t_j^0 \leq Y_i] \exp\{\boldsymbol{\beta}' \mathbf{X}_i\}$$
$$- n \sum_{j=1}^{d} p_\gamma(|\beta_j|) \tag{7.33}$$

with respect to $\lambda_1, \ldots, \lambda_N$. Here, $\mathbf{X}_j^*$ are the covariate values corresponding to each t_j^0. This maximization yields

$$\widehat{\lambda}_j = \frac{1}{\sum_{i \in D(t_j^0)} \exp\{\boldsymbol{\beta}' \mathbf{X}_i\}}, \quad j = 1, \ldots, N.$$

Now, plugging these $\widehat{\lambda}_j$ values into $\tilde{M}$ in Eq. (7.33) and a maximization with respect to $\boldsymbol{\beta}$ yields a regularized estimator of $\boldsymbol{\beta}$. The asymptotic properties of this procedure include consistency in selection and asymptotic normality of the sparsely estimated coefficients.

As an alternative method, Zhang and Lu [67] used the concept of Adaptive LASSO [69] in selecting variables in the Cox model. Here, the proposal was to minimize

$$-l_n + \gamma \sum_{j=1}^{d} |\beta_j| / |\tilde{\beta}_j|$$

where

$$l_n = \sum_{i=1}^{n} \delta_i \left[\boldsymbol{\beta}' \mathbf{X}_i - \ln \left\{ \sum_{j=1}^{n} I[Y_j \geq Y_i] \exp\{\boldsymbol{\beta}' \mathbf{X}_j\} \right\} \right],$$

γ is a penalty factor and $\tilde{\beta}$ is the unrestricted estimator of $\boldsymbol{\beta}$ obtained by maximizing the partial log likelihood l_n. This also yields a sparse estimator of the coefficient vector $\boldsymbol{\beta}$. The large sample properties of the resulting estimator of the coefficient vector are similar to those with Fan and Li [20].

7.2 Extending the Cox Proportional Hazards Model

7.2.1 Smoothed Estimate of the Baseline Hazard

The popularity of the Cox model may lie in the fact that to identify the effect of the covariates on the risk of failure it is not necessary to specify nor even to estimate the baseline hazard rate, that is, the complete model. This has an important consequence: the researchers who apply these methods to their particular interests do not usually pay attention to estimating the probability of failure. However, in many situations, a precise estimation of such probability becomes essential in order to acquire a satisfactory understanding of the phenomenon under study, since the baseline hazard is directly related to the time-course of the failure process.

The estimation of the baseline hazard function may be carried out by several methods, see Eq. (7.7) in Sect. 7.1.1, for instance. Likewise, our aim is to obtain an informative estimation for the baseline hazard, by a nonparametric procedure. Therefore, in this section, we develop flexible models within the proportional hazards framework. In particular, we consider smooth estimation of the baseline hazard by means of natural cubic splines [52].

The Cox model, as has been stated in Eqs. (7.2) and (7.3), can be expressed in terms of the cumulative hazard rate as

$$\Lambda(t, \mathbf{X}) = \Lambda_0(t) \exp(\boldsymbol{\beta}^\top \mathbf{X}). \tag{7.34}$$

The key idea consists of estimating λ by smoothing Λ. To do so, we first transform Eq. (7.34) by taking a natural logarithm, so we have

$$\ln[\Lambda(t, \mathbf{X})] = \ln[\Lambda_0(t)] + \boldsymbol{\beta}^\top \mathbf{X}. \tag{7.35}$$

Now we proceed to parametrize in a suitable way the first component in the sum of (7.35). In the particular case that the underlying distribution family is Weibull with characteristic life μ and shape parameter ρ, Eq. (7.35) takes the form

$$\ln[\Lambda(t, \mathbf{X})] = \rho \ln t - \rho \ln \mu + \boldsymbol{\beta}^\top \mathbf{X}, \tag{7.36}$$

which can be re-parametrized to

$$\ln[\Lambda(t, \mathbf{X})] = \gamma_0 + \gamma_1 x + \boldsymbol{\beta}^\top \mathbf{X}, \tag{7.37}$$

where $x = \ln t$. Thus, a linear function on the logarithmic scale of time is achieved.

In the general case of Eq. (7.35), $\ln \Lambda_0(t)$ is not necessarily related to the log-time by a linear function, and therefore we can express the general situation by means of a function ψ, getting the following expression for Eq. (7.35)

$$\ln[\Lambda(t, \mathbf{X})] = \psi(\ln t) + \boldsymbol{\beta}^\top \mathbf{X}, \tag{7.38}$$

where ψ denotes an unknown function defined over the log-scale time, i.e. $x = \ln t$. The following step is to think of flexible models for function ψ, and thus we consider an approach based on natural cubic splines as suggested in Royston and Parmar [52].

Cubic splines are the most popular spline functions. They are smooth functions with which to fit data, and when used for interpolation, they do not have the oscillatory behavior that is characteristic of high-degree polynomial interpolation (such as Lagrange interpolation, Hermite interpolation, etc.)

The idea behind computing a cubic spline is as follows. Let $\{x_0, x_1, \ldots, x_m\}$ be a set of m points (nodes). We can construct a cubic spline with m piecewise cubic polynomials between the given nodes by defining

$$s(x) = \begin{cases} s_0(x), & x \in [x_0, x_1]; \\ s_1(x), & x \in [x_1, x_2]; \\ \cdots\cdots\cdots \\ s_{m-1}(x), & x \in [x_{m-1}, x_m]. \end{cases} \tag{7.39}$$

To determine the m cubic polynomials $s_j(x) = a_j + b_j x + c_j x^2 + d_j x^3$ comprising s, we need to determine $4m$ unknown coefficients, so we need $4m$ constraints to solve the problem. Among other things, we require, for $j = 1, 2, \ldots, m - 1$,

(i) $s_{j-1}(x_j) = s_j(x_j)$;
(ii) $s'_{j-1}(x_j) = s'_j(x_j)$;
(iii) $s''_{j-1}(x_j) = s''_j(x_j)$.

In addition, if we make the second derivative of the 1st node and the last node zero then we have a *natural cubic spline*. That is, we also require that $s''(x_0) = s''(x_m) = 0$. This gives us $3(m - 1)$ constraints from (i)–(iii) plus 2 more constraints from the natural cubic spline condition. Thus, we have $3m - 1$ constraints leaving $m + 1$ degrees of freedom (df) to choose the coefficients. The number of degrees of freedom will determine the complexity of the curve, so we need to choose a number of $m + 1$ nodes in the range of the log survival times to complete the spline interpolation. If $m = 1$ is taken, i.e. only two points are selected $(df = 2)$, then the unknown function ψ is approximated by a linear function and the baseline distribution is assumed to be Weibull. The placement of the nodes is an issue, however, and as is explained in Royston and Parmar [52] and the references therein, optimal node selection does not appear critical for a good fit. We consider the extreme uncensored log survival times as the boundary nodes, that is x_0 is selected as the logarithm of the minimum of the observed uncensored survival times, and likewise x_m is the maximum. For internal node selection, we follow the recommendations of [52] and consider percentile-based positions. Thus, for a problem with $df = 3$ $(m = 2)$ the internal node is selected at the location of the median of the uncensored log survival times. If we consider a problem with $df = 4$, two internal nodes need to be determined, and then the 1st and the 3rd quartiles can be considered.

To summarize, imposing constraints (i)–(iii) on our spline function means that $m + 1$ unknown coefficients must be estimated together with the regression parameter vector $\boldsymbol{\beta}$. For convenience in the notation, let us denote by $\boldsymbol{\gamma}$ the $m + 1$ dimensional vector comprising the unknown parameters in the spline function. Thus, the model specified in (7.38) is approximated by

$$\ln[\Lambda(t, \mathbf{X})] = s(\ln t, \boldsymbol{\gamma}) + \boldsymbol{\beta}^\top \mathbf{X}. \tag{7.40}$$

A maximum likelihood procedure is conducted. Assume that we have n independent observations $\{(Y_i, \delta_i, \mathbf{X}_i), i = 1, 2, \ldots, n\}$, where the RCM is being considered once again; and $\mathbf{X}_i$ is a p dimensional vector of covariates defining the ith subject. Let L_i denote the likelihood contribution of the ith subject, so that the full likelihood function is obtained as

$$L = \prod_{i=1}^{n} L_i. \qquad (7.41)$$

For an uncensored observation, i.e. $\delta = 1$, we have

$$L_i = \exp\{s(\ln Y_i, \gamma) + \boldsymbol{\beta}^\top \mathbf{X}_i - \exp(s(\ln Y_i, \gamma) + \boldsymbol{\beta}^\top \mathbf{X}_i)\} \frac{ds}{dt}(\ln Y_i, \gamma). \qquad (7.42)$$

On the other hand, for an observation with $\delta = 0$, we obtain

$$L_i = \exp\{-\exp(s(\ln Y_i, \gamma) + \boldsymbol{\beta}^\top \mathbf{X}_i)\}. \qquad (7.43)$$

Example Insulating Fluid [46]

We have considered the data from Nelson [46], p. 277, which comprise the log times (in seconds) to breakdown of an insulating fluid at five voltage levels. We ignore here the highest voltage level (45 kV) since some left-censored data have been reported. Hence, we consider the log times to insulating fluid breakdown at 25, 30, 35 and 40 kV, respectively, that appear on page 278 of Nelson [46]. The experimental test reported six right-censored data when run at 25 kV and two at 30 kV. We have adopted the usual indicator variable coding methodology and constructed the following indicator variables:

$Volt30 = 1$, if the item is tested at voltage equal to 30 kV, 0 otherwise;

$Volt35 = 1$, if the item is tested at voltage equal to 35 kV, 0 otherwise;   (7.44)

$Volt40 = 1$, if the item is tested at voltage equal to 40 kV, 0 otherwise;

and hence, for an item tested at 25 kV, $Volt30 = Volt35 = Volt40 = 0$. The model (7.34) adopts the following form

$$\Lambda(t, \mathbf{X}) = \Lambda_0(t) \exp(\beta_1 Volt30 + \beta_2 Volt35 + \beta_3 Volt40). \qquad (7.45)$$

The Cox proportional hazards model has been fitted by using the functions contained in the package *survival* of R. According to the previous specifications, the model formula in the call to *coxph* specifies that time-to-breakdown depends on the level of voltage through the three indicator variables defined above. The initial fitting of the model yields the following results[1] reported by the function

---

[1] Same results may be achieved by using the function *strata* into the specification of function *coxph*.

Table 7.1 Cox proportional hazards model fit for the insulating fluid breakdown data

	Coef	Exp(Coef)	Se(Coef)	z	$Pr(>|z|)$
Voltage = 30 kV	1.0687	2.9116	0.4442	2.406	0.016140
Voltage = 35 kV	2.5131	12.3433	0.5688	4.419	9.94e-06
Voltage = 40 kV	2.5786	13.1782	0.7124	3.619	0.000295

Likelihood ratio test = 24.38 on 3 df, p-value = 2.086e−05
Wald test = 20.58 on 3 df, p-value = 0.0001289
Score (logrank) test = 24.72 on 3 df, p-value = 1.771e−05

Table 7.2 Spline smoother results for the insulating fluid breakdown data

Newton–Raphson maximization

Number of iterations: 9

Function value: −353.7648

Estimates:

	Estimate	Gradient
Coef. for spline (1)	−5.7480422140	−1.818989e−06
Coef. for spline (2)	0.4892591763	−2.387424e−06
Coef. for spline (3)	−0.0001843661	−3.410605e−07
Voltage = 30 kV	1.3344481665	0.000000e+00
Voltage = 35 kV	2.7773802570	−1.477929e−06
Voltage = 40 kV	2.8436290388	5.684342e−08

summary which lead us to admit that the Cox model seems to be, at least in principle, reasonable for these data with the voltage factor, although analysis of residuals to investigate lack-of-fit from the model is presented later.

Next, we proceed to obtain a smoothed estimate of the cumulative hazard function by means of the spline procedure explained previously. Since the size of the sample is small, only 48 data are available, we formulate the model (7.38), by considering a unique node in the definition of the spline, that is, only two piece-wise cubic polynomials are composed to fit the data. The restrictions involved by the natural spline definition lead us to have 3 parameters to be estimated for the spline part of the model, plus 3 parameters corresponding to the coefficients of the indicator variables describing the covariate categories.

We have constructed the log likelihood function as explained above, and we have used a maximization procedure to find estimations for the parameters in the model. In particular, we have used the functions included in the package *maxLik* [60] of the program R. Several maximization functions have been used and the final estimations are displayed below.

The three first coefficients presented in Table 7.2 give estimations of coefficients required to construct the spline function to model the log cumulative baseline hazard function. The other three values are the estimations of the coefficients for the function on the covariate Voltage. As it can be noted, these values are similar to the ones obtained by conducting a standard Cox model fitting. Graphical results are presented on Fig. 7.4. The different graphs also show the fit

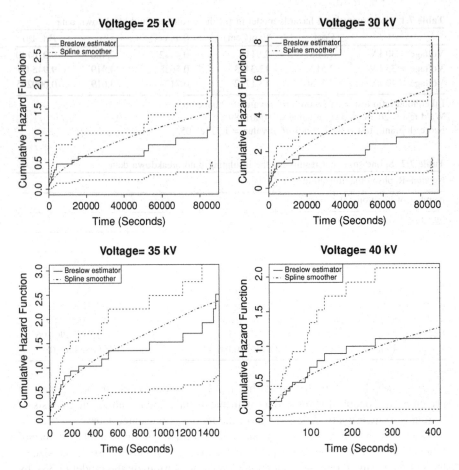

Fig. 7.4 Insulating fluid example

to a Cox model. In the top-left panel, we present the estimated baseline cumulative hazard function and confidence bands, based on the Breslow estimator, which has been obtained through the *survival* package of R. The dashed line represents the estimated cumulative hazard function based on the spline approach. The other panels show the same information for the population with the respective levels of the covariate indicated on top of the corresponding graph. As can be appreciated, the results for the group of subjects tested at a voltage level of 30 kV do not match. This may suggest a spline function with a higher degree of complexity, i.e. we could try to formulate a model with two nodes instead of the current model with only one node located at the median value of the log lifetimes. On the other hand, in the regression model fitted under the proportional hazards assumption and reported in Table 7.1, the coefficient corresponding to the group of subjects tested at a voltage level of 30 kV has the highest p-value associated with the individual

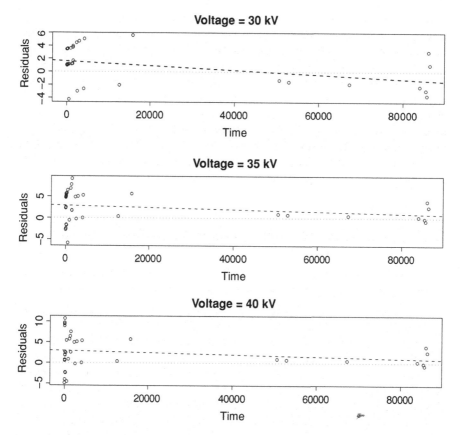

Fig. 7.5 Plots of scaled Schoenfeld residuals against time for voltage covariate in a Cox model fit

Table 7.3 Global and individual tests for proportionality, based on Schoenfeld residuals

	Pearson correlation	χ^2	p-value
Factor (voltage) 30	−0.412	5.328	0.021
Factor (voltage) 35	−0.216	1.489	0.222
Factor (voltage) 40	−0.183	0.987	0.321
Global	NA	5.328	0.149

significance tests. Specifically, the result of this test suggests that this coefficient is not significant at 1% of significance level, so the corresponding plot in Fig. 7.4 should be interpreted with caution.

We have investigated the underlying proportional hazards assumption by exploring the Schoenfeld residuals. The R default for checking the proportional hazard assumption is a formal test and plot of scaled Schoenfeld residuals. This is done by calling to the function *cox.zph* of the package *survival*. Figure 7.5 plots the scaled Schoenfeld residuals versus time for each covariate, together with a scatterplot smooth and a line with zero slope. When the proportional hazards assumption holds, a relatively straight horizontal line is expected. The plot for *Voltage* = 30 kV is showing some negative linear correlation with time, and therefore it is alerting on some evidence of deviation from the proportional assumption. The standard tests for proportionality using the survival package of R rely on some specific modeling of the dependence of time, that is, it is considered that the coefficients $\beta(t) = \beta + g(t)$ with $g(t) = t$, $\log(t)$ or the rank of the event times. In Table 7.3, the results of these tests, for $g(t) = t$, suggest a lack-of-fit due to nonproportional hazards for covariate Voltage = 30 kV. The results presented in the table give estimations for the Pearson correlation coefficients between residuals and $g(t) = t$ for each covariate level.

7.2.2 The Generalized Additive Proportional Hazards Model

In this section, generalizations of the Cox model to a completely Nonparametric model are introduced. In this case, an arbitrary covariate effect of the form

$$\lambda(t, \mathbf{X}) = \lambda_0(t) \exp[g(\mathbf{X})], \tag{7.46}$$

is considered, where g is an unspecified smooth function of the covariate vector, $\mathbf{X}$.

Although the model in (7.46) is the most general proportional hazards model, it becomes difficult to estimate $g(\mathbf{X})$ in the case of a high dimensionality of vector $\mathbf{X}$ because of the sparseness of data. This problem is known as the *curse of dimensionality*. To overcome the problem, dimension reduction methods are required and in this context, the additive proportional hazards model offers a convenient alternative. If g in Eq. (7.46) is assumed to be additive, the model can be expressed as

$$\lambda(t, \mathbf{X}) = \lambda_0(t) \exp\left[\sum_{r=1}^{p} g_r(X_r)\right], \tag{7.47}$$

where $g_r(X_r)$ are unspecified smooth functions. The problem of estimating these smooth functions has been addressed by many authors. The methods for estimating this function that are considered in this section can be classified into two groups:

- *Partial likelihood methods*;
- *Time transformation methods*.

7.2.2.1 Partial Likelihood Methods

Let us consider again a censored sample $\{(Y_i, \delta_i, \mathbf{X}_i); i = 1, 2, \ldots, n\}$, where, as always, $Y_i = \min\{T_i, C_i\}$, T_i and C_i being, respectively, the failure time and the censoring time of the ith sample element; and $\delta_i = I[Y_i = T_i]$, for $i = 1, 2, \ldots, n$. The partial likelihood method based on this sample computes the log partial likelihood

$$PL(g_1, g_2, \ldots, g_p) = \prod_{i=1}^{n} \left[\frac{e^{\sum_{r=1}^{p} g_r(X_{r,i})}}{\sum_{j=1}^{n} D_j(Y_i) e^{\sum_{r=1}^{p} g_r(X_{r,j})}} \right]^{\delta_i} \tag{7.48}$$

where $D_j(Y_i) = I[Y_j \geq Y_i]$ denotes the at-risk process, as in Sect. 7.1.1. Observe the parallelism between this equation and Eq. (7.6). Maximizing the logarithm of the above expression yields estimates $\widehat{g}_1, \ldots, \widehat{g}_p$; however, the computations involved can be very time-consuming. Localized versions of this partial likelihood can be introduced [18, 21].

Let us explain the method in the case of $p = 1$; generalizations to higher dimensions can easily be developed. The logarithm of the partial likelihood for $p = 1$ is obtained as

$$\ln PL(g) = \sum_{i=1}^{n} \delta_i \left[g(X_i) - \ln \left(\sum_{j=1}^{n} D_j(Y_i) e^{g(X_j)} \right) \right]. \tag{7.49}$$

If the function g is smooth in the sense that the dth order derivative exists, the Taylor expansion could be considered, and we could then approximate g in the above expression by a polynomial of m degree given by

$$\begin{aligned} g(X) \approx p(X) &= g(x) + g'(x)(X - x) + \cdots + \frac{d^m g(x)}{dx^m} \frac{(X - x)^m}{m!} \\ &= \beta_0 + \beta_1(X - x) + \cdots + \beta_m(X - x)^m, \end{aligned} \tag{7.50}$$

for X in a neighborhood of x. Thus, the expression in (7.49) is transformed to

$$\ln PL(g) \approx \sum_{i=1}^{n} \delta_i \left[p(X_i) - \ln \left(\sum_{j=1}^{n} D_j(Y_i) e^{p(X_j)} \right) w_h(X_j - x) \right] w_h(X_i - x), \tag{7.51}$$

where $w_h(\cdot) = w(\cdot/h)/h$ and w is a kernel density function and $h > 0$ is a bandwidth parameter that controls the size of the local neighborhood. The kernel function determines the weights assigned to the observations around x. Let us denote $g^{(r)} = \frac{d^r g(x)}{dx^r}$, for $r = 1, 2, \ldots, m$. Now, the objective is to maximize the above expression and obtain estimates

$$\widehat{g}^{(r)} = r! \widehat{\beta}_r,$$

for $r = 0, 1, \ldots, m$. It is noticeable that $g(x) = \beta_0$ is not directly estimable from (7.51), since the corresponding term is canceled in the expression. An estimate of

$g(x)$ can be found by numerical integration, for which the literature recommends the trapezoidal rule.

To estimate the whole function g, the local partial likelihood (7.51) must be applied at a large number of locations x. The drawback of these methods is that the computations involved can be very time-consuming. Furthermore, as pointed out, it must be highlighted that the function g is not estimated directly with this procedure. Instead, a numerical integration procedure must be conducted to obtain it from its derivative. Therefore, an alternative method for estimating g is proposed in the next section.

7.2.2.2 Time Transformation Methods

The procedure we demonstrate in this subsection was suggested by Kvaløy [38]. The basic idea is to transform the estimation problem for the nonparametric Cox model to a problem of exponential regression. Time transformation models for Cox regression have been considered by, for example, Gentleman and Crowley [24] and Clayton and Cuzik [12].

Let $\{(Y_i, \delta_i, \mathbf{X}_i); i = 1, 2, \ldots, n\}$, be a sample according to the RCM model, with $\mathbf{X}_i = (X_{r,i};\ r = 1, \ldots, p)$ a vector of covariates describing the ith sample element, and δ_i the censoring indicator, for $i = 1, 2, \ldots, n$.

Algorithm 7.1 (*The General Time Transformation Algorithm (GTTA)*)

Step 1. Find an initial estimate $\widehat{\Lambda}_0(t)$.

Step 2. Transform $Y_1, Y_2, \ldots, Y_n$ to $\widehat{\Lambda}_0(Y_1), \widehat{\Lambda}_0(Y_2), \ldots, \widehat{\Lambda}_0(Y_n)$.

Step 3. Estimate $g(\mathbf{X})$ from $\left(\widehat{\Lambda}_0(Y_1), \delta_1, \mathbf{X}_1\right), \ldots, \left(\widehat{\Lambda}_0(Y_n), \delta_n, \mathbf{X}_n\right)$.

Step 4. Find a new estimate of $\Lambda_0(t)$ from $(Y_1, \delta_1, \widehat{g}(\mathbf{X}_1)), \ldots, (Y_n, \delta_n, \widehat{g}(\mathbf{X}_n))$.

Step 5. Repeat Step 2–Step 4 until convergence.

The initial estimate of Step 1 is obtained by fitting a standard Cox model, that is, assuming that the g_r are linear functions, for $r = 1, 2, \ldots, p$. The algorithm is based on the fact that if a random variable T has hazard function $\lambda(\cdot)$ then $\Lambda(T) = \int_0^T \lambda(u)\, du$ is exponentially distributed with parameter 1. Therefore, if $\Lambda(T, \mathbf{X}) = \Lambda_0(T)g(\mathbf{X})$ then $\Lambda_0(T)$ has an exponential distribution with parameter $g(\mathbf{X})$ conditional on $\mathbf{X}$. This idea explains Step 3 of the above GTTA where $g(\mathbf{X})$ is estimated using a method for exponential regression. Although other methods are available in the literature, we will focus on the *covariate order method* introduced by Kvaløy [38] and subsequently studied in Kvaløy and Lindqvist [39, 40]. This method can be outlined as follows.

Let (V_i, δ_i, X_i) for $i = 1, 2, \ldots, n$ be a censored random sample from the model (V, δ, X), where $V = \min\{Z, C\}$, $\delta = I[Z \le C]$, and X is a single continuous covariate distributed according to a $pdf f(x)$. Given $X = x$, the variables Z and C are stochastically independent, and moreover Z is exponentially distributed with hazard parameter $\lambda(x)$. Note that a different notation is now introduced, Z, to distinguish it from the target lifetime random variable T, and V for the corresponding censored version. The random variable C is distributed according to an unspecified density function, which may depend on x.

The steps to carry out the covariate order method for exponential regression are summarized below.

Algorithm 7.2 (*The Covariate Order Algorithm (COA)*) [38]

Step 1. Arrange the observations $(V_1, \delta_1, X_1), (V_2, \delta_2, X_2), \ldots, (V_n, \delta_n, X_n)$ such that $X_1 \leq X_2 \leq \ldots \leq X_n$

Step 2. For convenience, compute the scaled observation times V_i/n, for $i = 1, 2, \ldots, n$ and define $S_k = \sum_{i=1}^{k} V_i/n$, for $k = 1, 2, \ldots, n$.

Step 3. Construct an artificial point process for which it is considered that an event occurs at the time S_k if and only if $\delta_k = 1$. If $m = \sum_{i=1}^{n} \delta_i$ then define $S_1^*, S_2^*, \ldots, S_m^*$ as the arrival times of this process.

Step 4. Let $\rho(s)$ denote the conditional intensity of the process $S_1^*, S_2^*, \ldots, S_m^*$, making it easy to see that $\rho(s) = n\lambda(X_k)$ for $s \in (S_{k-1}, S_k]$.

Step 5. Construct a nonparametric estimate of the conditional intensity function of the process $S_1^*, S_2^*, \ldots, S_m^*$. For example, consider a kernel type estimate as given by

$$\widehat{\rho}(s) = \frac{1}{h} \sum_{j=1}^{m} k\left(\frac{s - S_j^*}{h}\right),$$

where $k(\cdot)$ is a kernel function, that is, a compactly supported, bounded, symmetric *pdf*; and h is a smoothing parameter.

Step 6. Estimate the *correspondence function*, that is, the relationship between the covariate values and the time-scale axis of the process $S_1^*, S_2^*, \ldots, S_m^*$. This can be any function $s(x)$ which relates the covariate value x to a representative point $s = s(x)$ on the axis where the artificial processes $S_1, S_2, \ldots$ and $S_1^*, S_2^*, \ldots$ develop. The easiest such function is perhaps the step function $\widetilde{s}(x) = S_k$ for $x \in [X_k, X_{k+1})$. Alternative functions are obtained by linear interpolation between points $(X_1, S_1), (X_2, S_2), \ldots, (X_n, S_n)$, or by using more sophisticated smoothing techniques.

Step 7. An estimator of $\lambda(x)$ can thus be given by

$$\widehat{\lambda}(x) = \widehat{\rho}(\widetilde{s}(x))/n.$$

In the above computation of $\widehat{\rho}$ at $\widetilde{s}(x)$, using the formula of Step 5, th h of Step 5 may well be chosen to depend specifically on x, as h_x. In this case the h used in Step 5 would be $h = \widetilde{s}(x + h_x/2) - \widetilde{s}(x - h_x/2)$. It has been shown in Kvaløy and Lindqvist [40] that under regularity conditions the $\widehat{\lambda}(x)$ converges uniformly in x to the true underlying $\lambda(x)$.

Figure 7.6 illustrates steps 2 and 3 in the Covariate Order Algorithm.

This algorithm is used in Kvaløy [38] in Step 3 of the GTTA in order to estimate the component functions g_r in (7.47). Below, we present the resulting algorithm basically following Kvaløy [38].

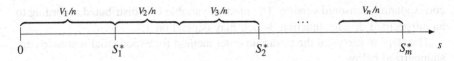

Fig. 7.6 Construction of the artificial process in the Covariate Order Algorithm. Here, $\delta_1 = \delta_3 = \delta_n = 1$, while $\delta_2 = 0$

Note that since the COA can only work with a one-dimensional covariate, we need to update one function $\widehat{g}_r$ at a time. The basic idea of the algorithm is that, since for a given vector of covariates $\mathbf{X}$, $\Lambda(T, \mathbf{X})$ follows an exponential distribution with rate equal to 1, $\Lambda_0(T) \exp[g_1(X_1) + \cdots + g_{r-1}(X_{r-1}) + g_{r+1}(X_{r+1}) + \cdots + g_p(X_p)]$ follows an exponential distribution with rate parameter $\exp[g_r(X_r)]$, for each $r = 1, 2, \ldots, p$. The algorithm below considers a slight modification of the procedure given in Kvaløy [38].

Algorithm 7.3 (*The Time Transformation Method with Covariate Order for Exponential Regression*)   Set $\widehat{g}_r^{(0)} = X_r$ for $r = 1, 2, \ldots, p$ as initial values, and let $k = 0$.

Step 1. Fit a standard Cox model to the original data, that is, consider the model

$$\lambda(t, X_1, X_2, \ldots, X_p) = \lambda_0(t) \exp\left(\beta_1 X_1 + \beta_2 X_2 + \cdots + \beta_p X_p\right)$$

and obtain estimates $\widehat{\Lambda}_0^{(0)}$ and $\widehat{\boldsymbol{\beta}}^{(0)} = \left(\widehat{\beta}_1, \widehat{\beta}_2, \ldots, \widehat{\beta}_p\right)^{\top}$.

Step 2. Do $k = k + 1$. Construct the $n \times p$ dimensional matrix whose entries are

$$\Lambda_{(-r)}^{(k)}(Y_i) = \widehat{\Lambda}_0^{(k-1)}(Y_i) \exp\left[g_1^{(k-1)}(X_1) + \cdots + g_{r-1}^{(k-1)}(X_{r-1}) + \right.$$
$$\left. + g_{r+1}^{(k-1)}(X_{r+1}) + \ldots + g_p^{(k-1)}(X_p)\right]$$

for $i = 1, 2, \ldots, n$ and $r = 1, 2, \ldots, p$.

Step 3. Using the method described in the COA, solve p exponential regression problems from each sample of the type $\left\{\left(\Lambda_{(-r)}^{(k)}(Y_i), \delta_i, X_{r,i}\right);\right.$ for $i = 1, 2, \ldots, n\}$. This produces a vector of estimated functions $\widetilde{g}_1^{(k)}(x_1), \widetilde{g}_2^{(k)}(x_2), \ldots, \widetilde{g}_p^{(k)}(x_p)$.

Step 4. Define the centered functions $\widehat{g}_r^{(k)}(x_r) = \widetilde{g}_r^{(k)}(x_r) - \sum_{i=1}^n \widetilde{g}_r^{(k)}(X_{r,i})/n$, for each $r = 1, 2, \ldots, p$.

Step 5. Obtain the Breslow estimate of the cumulative baseline hazard function from

$$\widehat{\Lambda}_0^{(k)}(t) = \sum_{i:Y_i \leq t} \frac{\delta_i}{\sum_{j=1}^n D_j(Y_i) \exp\left[\sum_{r=1}^p \widehat{g}_r^{(k)}(X_{r,j})\right]},$$

or fit a standard Cox model using the functions $\widehat{g}_r^{(k)}$ to obtain a new $\widehat{\Lambda}_0^{(k)}$.

Step 6. Repeat 2–5 until convergence.

Remarks 7.2

- Step 3 of the algorithm is different from the one in the algorithm presented in [38], where a backfitting method is considered.
- Step 4 is introduced in order to make the model identifiable. Otherwise, the model is only identifiable up to a constant, since it is easy to check that $\lambda_0(t) \exp[g(\mathbf{X})] = c\lambda_0(t) \exp[g(\mathbf{X}) - \log c]$.
- Stratified models of the form $\lambda_{0q}(t) \exp[g(\mathbf{X})]$, $q = 1, \ldots, s$, where s is the number of strata, can also be fitted with a simple modification of the algorithm. The only change needed in the algorithm is to estimate a $\widehat{\Lambda}_{0q}(t)$ for each stratum in Steps 1 and 5 and transform each Y_i in Step 2 according to the stratum to which it belongs.

Example Tire Reliability Analysis

Krivtsov et al. [37] present data from an empirical study regarding root-causes of a certain kind of automobile tire failure. The data were obtained from a laboratory test that was developed to duplicate field failures. Six different parameters related to tire geometry and physical properties were selected as explanatory variables that

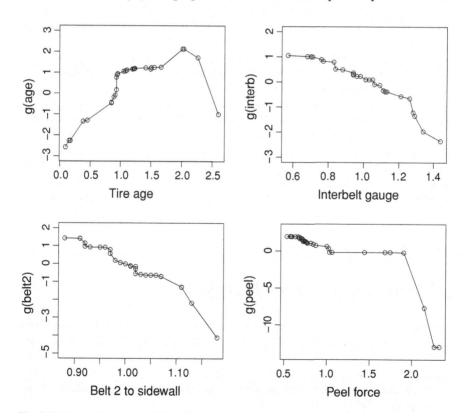

Fig. 7.7 Plots of the estimated functions $g(\cdot)$ for the tire data from Krivtsov et al. [37]

potentially affect a tire's life on test. The data consist of lifetimes for 34 tires, with as many as 23 of these being right-censored. It was, however, still possible to conclude significant influence for several of the explanatory variables in a Cox regression analysis.

The final model chosen by Krivtsov et al. [37] involves four variables, one of which is an interaction between two of the original variables. In order to illustrate the use of an additive model, we have instead chosen another set of four explanatory variables which together give an adequate description of the data. These are *Tire age* (age), *Interbelt gauge* (interb), *Belt 2 to sidewall* (belt2) and *Peel force* (peel). We refer to the article for more detailed explanations of these variables.

The model we would like to fit has the hazard rate:

$$\lambda_0(t) \exp[g(\text{age}) + g(\text{interb}) + g(\text{belt2}) + g(\text{peel})].$$

Figure 7.7 shows the result using Algorithm 7.3. Note that due to edge effects, and also few observations near the ends of the intervals where the covariates take their values, the estimated functions are believed to be more reliable in the middle parts of their domains (although an edge correction based on reflection, suggested by Kvaløy [38], has been used). The kernel used is the Epanechnikov kernel, and the window sizes h used in Step 5 of the COA are chosen using the following values h_x (see Step 7): 1.7 (age), 0.7 (interb), 0.4 (belt2) and 2.7 (peel). The final estimates of the $g(\cdot)$ are furthermore adjusted by a multiplicative constant in order that a Cox regression using the final $g(\cdot)$ as covariates gives all coefficients equal to 1 (which is otherwise not guaranteed by the algorithm).

The graphs indicate that an ordinary Cox model with linear functions $g(\cdot)$ could be appropriate for all variables, except possibly Peel force. This covariate is, on the other hand, the one which gives the least significant coefficient β in an ordinary Cox regression analysis using the four variables. More specifically, the estimated coefficients of a Cox analysis obtained from the package *survival* in R, are (with *p*-value appearing in parentheses), $\beta(\text{age}) = 2.30$ (0.053), $\beta(\text{interb}) = -3.49$ (0.054). $\beta(\text{belt}) = -21.48$ (0.025), $\beta(\text{peel}) = -6.50$ (0.110). We note that these coefficients correspond well to the average slopes of the corresponding curves in Fig. 7.7.

7.3 Nonparametric Hazard Regression

The task of smoothing in hazard regression problems can be tackled using two different approaches, depending on the way the problem is dealt with when covariates are not present. This leads us to two families of estimators for the hazard rate: the so-called "external" estimators and "internal" estimators, as they are classified in Nielsen and Linton [50] following the language introduced by Jones et al. [29].

- *"External" Estimators* Given the definition of the hazard function as the ratio of the density and the survival function $\lambda(t) = f(t)/S(t)$, this family of estimators results from replacing f and S with nonparametric estimators $\widehat{f}$ and $\widehat{S}$, respectively.
- *"Internal" Estimators* The second family of estimators comes from smoothing the increments of a nonparametric estimate of the cumulative hazard function $\widehat{\Lambda}$. For example, consider the following kernel estimator for the hazard function

$$\widehat{\lambda}(t) = \int \omega_b(t-u)\,\mathrm{d}\widehat{\Lambda}(u),$$

where $b > 0$ is a bandwidth parameter and ω_b a kernel function that depends on the bandwidth.

7.3.1 Local Linear Smoothing of the Hazard Rate

The methods used in this section will focus on the first approach, thus giving local (constant and linear) estimators for the conditional density and survival functions. The methods that we develop below have been suggested in Spierdijk [56] and can be considered a simplification of the local linear estimator introduced by Nielsen [49] in the context of hazard regression.

As explained in Sect. 7.1.1, we start with a sample of independent observations of size n in the form $(Y_i, \delta_i, \mathbf{X}_i)$, $i = 1, 2, \ldots, n$, where Y_i are right-censored lifetimes, which we assume ordered. Under the classical random censoring model, independent random variables T and C exist, denoting, respectively, the lifetime and the censoring time. Therefore, for each $i = 1, 2, \ldots, n$, the corresponding sample value is an observation of the random variable $Y = \min\{T, C\}$, and hence, $\delta_i = I[Y_i = T_i]$. For convenience, we are going to formulate the hazard function corresponding to T, that is $\lambda(t; \mathbf{X})$, in terms of the observed data. Let $H(t; \mathbf{X}) = (1 - F(t; \mathbf{X}))(1 - G(t; \mathbf{X}))$ denote the survival function corresponding to Y, where G is the distribution function of C. Thus, according to Eq. (7.1), if $G(t; \mathbf{X}) < 1$, we have

$$\lambda(t; \mathbf{X}) = \frac{f(t; \mathbf{X})}{1 - F(t; \mathbf{X})} \frac{1 - G(t; \mathbf{X})}{1 - G(t; \mathbf{X})} = \frac{f(t; \mathbf{X})(1 - G(t; \mathbf{X}))}{H(t; \mathbf{X})}. \tag{7.52}$$

If we concentrate on the joint distribution of the random vector (Y, δ), we can argue for $\delta = 1$ that

$$\begin{aligned} h(t, \delta = 1) &= \lim_{\Delta \to 0} P\{t < Y \le t + \Delta; \delta = 1\} \\ &= \lim_{\Delta \to 0} P\{t < T \le t + \Delta; C > t + \Delta\}, \end{aligned} \tag{7.53}$$

since T and C are assumed to be independent, the expression above leads to

$$h(t, \delta = 1) = \lim_{\Delta \to 0} P\{t < T \le t + \Delta\} \lim_{\Delta \to 0} P\{C > t + \Delta\} = f(t)(1 - G(t)),$$

and then, Eq. (7.52) reduces to

$$\lambda(t; \mathbf{X}) = \frac{h(t, \delta = 1; \mathbf{X})}{H(t; \mathbf{X})}. \tag{7.54}$$

Therefore, to obtain a nonparametric estimate of the conditional hazard rate, we replace $h(t, \delta = 1; \mathbf{X})$ and $H(t; \mathbf{X})$ with the corresponding nonparametric estimates. To do this, we consider smooth estimates for each of them and construct $\widehat{h}(t, \delta = 1; \mathbf{X})$ and $\widehat{H}(t; \mathbf{X})$ from the observed dataset.

For simplicity, we assume that the covariate vector is a scalar. The generalization to higher dimensions is easy. To estimate h_1, we only have to take into account the uncensored observations, whereas for the estimator of the function in the denominator H, we make use of all the observations regardless of whether they are censored or not.

Let us define the following kernel functions

$$w_{b_1}(t) = w(t/b_1)/b_1, \quad k_{b_2}(x) = k(x/b_2)/b_2,$$

where $w(\cdot)$ and $k(\cdot)$ are density functions, and b_1 and b_2 are suitable bandwidth parameters. We consider first a nonparametric estimate of $h(t, \delta = 1)$ without consideration of the covariate effect and then we smooth such an estimate. As it was explained in Chap. 1, we can define the following estimator for h

$$\widehat{h}(t, \delta = 1) = \frac{1}{n} \sum_{i=1}^{n} w_{b_1}(t - T_i)\delta_i, \tag{7.55}$$

but we can also argue by conditioning on the event $\{\delta = 1\}$.

$$h(t, \delta = 1) = \lim_{\Delta \to 0} P\{t < Y \le t + \Delta | \delta = 1\} P\{\delta = 1\} = h_1(t) P\{\delta = 1\} \tag{7.56}$$

and then define estimators for the two factors in this last expression. A natural estimator of $P\{\delta = 1\}$ is given by n_1/n, where n_1 is the number of uncensored observations in the sample. To estimate h_1, it is sufficient to consider the uncensored subsample, so, we can define as above

$$\widehat{h}_1(t) = \frac{1}{n_1} \sum_{i \in U} w_{b_1}(t - T_i), \tag{7.57}$$

where $U \subseteq \{1, 2, \ldots, \}$ denotes the set of indices that correspond to uncensored observations. The estimate in (7.57) applies a weight of $1/n_1$ to each (uncensored) sample observation. This is the kernel estimator of the density h_1 in the case of nonexistence of the covariates. Let us now consider a smoothed version of the estimator by assigning local weights to the observations in terms of the covariate component. So, we define the following

$$\widehat{h}_1(t;x) = \sum_{i\in U} k^*_{b_2,i}(x) w_{b_1}(t - T_i) \tag{7.58}$$

For simplicity, we consider once again the Nadaraya-Watson weights

$$k^*_{b_2,i}(x) = \frac{k_{b_2}(x - X_i)}{\sum_{j\in U} k_{b_2}(x - X_j)},$$

$$= \frac{k\left(\frac{x-X_i}{b_2}\right)}{\sum_{j\in U} k\left(\frac{x-X_j}{b_2}\right)}, \tag{7.59}$$

for all $i \in U$. One of the advantages of the estimator constructed by means of (7.58) and (7.59) is that it can be obtained using existing software and specifically with functions implemented in the package *np* [27] of R as we will see in the example below.

Of course, we could have conducted the smoothing procedure directly from the estimator defined in (7.55).

The resulting estimator of the conditional hazard function is closely related to the one suggested in Nielsen and Linton [50], but now it is not considered that the kernel function for smoothing in the time variable w is necessarily the same that for the covariate k, and, in the same way, the smoothing parameters b_1 and b_2 are allowed to be different.

The weights could also have been chosen based on a local linear smoothing procedure as developed in Spierdijk [56]. One of the advantages is that the boundary effect can be better overcome using the local linear approach. In such a case, a weighted least squares regression of $w_{b_1}(t - T_i)$ on $(x - X_i)$, with weights given by $k_{b_2}(x - X_i)$, is considered. Thus, the following minimization problem must be solved

$$\min_{\alpha,\beta} \sum_{i\in U} \{w_{b_1}(t - T_i) - \alpha - \beta(x - X_i)\}^2 k_{b_2}(x - X_i). \tag{7.60}$$

Therefore, the local linear estimator is the intercept term, that is $\widehat{\alpha}$, which is obtained as

$$\widehat{\alpha} = (1,0) \begin{pmatrix} v^0(x) & v^1(x) \\ v^1(x) & v^2(x) \end{pmatrix}^{-1} \begin{pmatrix} u^0(x) \\ u^1(x) \end{pmatrix} \tag{7.61}$$

where

$$v^r(x) = \sum_{i\in U} k_{b_2}(x - X_i)(x - X_i)^r, \quad r = 0, 1, 2;$$

and

$$u^r(x) = \sum_{i \in U} k_{b_2}(x - X_i)(x - X_i)^r w_{b_1}(t - T_i), \quad r = 0, 1.$$

This estimator was defined by Fan et al. [22]. It can be written in accordance with the notation in Eq. (7.58) by defining

$$k_{b_2,i}^*(x) = (1, 0) \begin{pmatrix} v^0(x) & v^1(x) \\ v^1(x) & v^2(x) \end{pmatrix}^{-1} \begin{pmatrix} k_{b_2}(x - X_i) \\ k_{b_2}(x - X_i)(x - X_i) \end{pmatrix}. \quad (7.62)$$

For an estimation of the function H, we proceed in a similar way. We define

$$W(t) = \int_{-\infty}^{t} w(u) \, du, \quad W_{b_1} = W(t/b_1). \quad (7.63)$$

So the corresponding estimator for the survival function H can be obtained by

$$\widehat{H}(t; x) = \sum_{i=1}^{n} k_{b_2,i}^*(x) W_{b_1}(t - T_i). \quad (7.64)$$

Note that all the observations $\{(Y_i, X_i); i = 1, 2, \ldots, n\}$ are now used in the estimation procedure, regardless of whether they are censored or not. For the weights $k_{b_2,i}^*(x)$ one of the two options explained above, Nadaraya–Watson or, for example, local linear weights, could be considered.

The local linear estimator considered in this section supposes a little modification of the one defined in Nielsen [49]. In the paper, the author carries out a multivariate local linear fitting based on least squares for the whole hazard regression problem. He shows that the estimator suggested in Nielsen and Linton [50], which is equivalent to the one obtained here by considering the Nadaraya–Watson weights, is a local constant estimator. In addition, Nielsen [49] obtains results concerning the pointwise convergence of the local (constant and linear) estimators. In our case, the local linear approach is developed only for the smoothing which concerns the covariate arguments, whereas for the time variable, the corresponding kernel density estimator, in the numerator, and kernel estimator of the survival function, in the denominator, are used.

Example A Water Supply Network in the South of Spain[2]

As it was mentioned above, for constructing the examples in this section we will make use of certain functions contained in the *np* package of R, where, among other things, a variety of nonparametric and semi-parametric kernel-based estimators are implemented. One of the most interesting features of this package is

---

[2] The authors express their gratitude to Prof. Carrión (Universidad Politécnica de Valencia, Spain) for providing the data in this example.

that it allows to handle sample sets that may contain simultaneously categorical and continuous datatypes, which is rather useful in practice.

The data that we analyze here are a subsample of the dataset previously considered in [8]. These data consist of failure times registered in pipes of a water supply network in a city in the Mediterranean Coast of Spain. Each datum in the original sample corresponds to a "pipe section" and registers contain information concerning installation year; section length; section diameter; pipe material; traffic conditions. The effect of the Length covariate (measured in meters) over the hazard function will be the only one analyzed in this example. More specifically, given a pipe section of a particular length, we wish to nonparametrically estimate the conditional hazard function of the time until breakdown.

We have selected the items whose length is lower than 200 m, which represent more than 98% of the items in the original dataset. We have discarded the remaining pipe sections because we do not consider them as representative of the population with respect to the covariate Length in the sense that there is less than a 2% of observations in an interval ranging from 200 to 1,870 m. Thus, we are working with 25971 registers in our dataset of which only 963 correspond to observed failure times. We do not distinguish between "natural" failures, due to aging of the pipe and failures due to "accidents" of any type. Thus, the censoring rate of the data is about 96%. Besides, although the observation window is 2000–2006, many of the pipes were installed very far in the past, so, in addition to the right-censoring, left-truncation is present in the dataset. Nevertheless, for our purposes in this example, the left-truncation feature will be ignored, so we will consider just the right-censoring in the dataset.

Following the explanations above, the problem of hazard regression is split in two parts. On the one side, we estimate the conditional density function of the time-to-failure given the covariate (Length) using just the uncensored information; and, on the other side, we need to estimate the conditional survival function of the time-to-failure given the covariate, on the basis of the whole dataset regardless of whether each observation is censored or not.

To do this, we make use respectively of the functions *npcdens* and *npcdist*. The approach is based on the work of Hall et al. [26] who use the so-called "generalized product kernels" capable of dealing with continuous and categorical datasets for estimating conditional probability densities. In this first example, we consider only continuous datatypes.

We use the default *fixed* bandwidths defined in the *npcdens* and *npcdist* functions. Also we use second-order Gaussian kernels (default). Finally, we modify the default argument that controls the method to compute the bandwidth, which is likelihood cross-validation. Instead, we consider the "normal-reference" argument, which is no other than the rule-of-thumb bandwidth computed by means of the standard formula

$$b_j = 1.06 \times \sigma_j \times n^{-1/(2 \times P + l)},$$

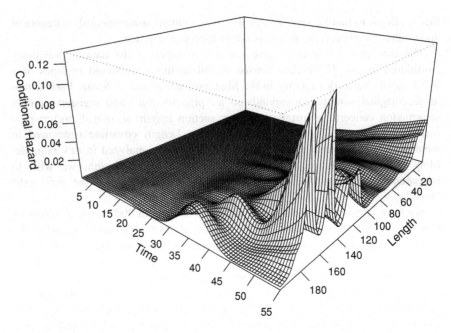

Fig. 7.8 Pipes breakdown data example

Table 7.4 Conditional density and distribution estimation for breakdown water supply data

Conditional Density Data: 963 training points, and 10000 evaluation points, in 2 variable(s)
(1 dependent variable(s), and 1 explanatory variable(s))
Time-to-breakdown
Dep. Var. Bandwidth(s): 3.248289
Length
Exp. Var. Bandwidth(s): 13.06992
Bandwidth Type: Fixed
Log Likelihood: −43667.56
Continuous Kernel Type: Second-Order Gaussian
No. Continuous Explanatory Vars.: 1
No. Continuous Dependent Vars.: 1
Conditional distribution data: 25971 training points, and 10000 evaluation points, in 2
     variable(s)
(1 dependent variable(s), and 1 explanatory variable(s))
Time-to-breakdown
Dep. Var. Bandwidth(s): 2.604847
Length
Exp. Var. Bandwidth(s): 4.497222
Bandwidth Type: Fixed
Continuous Kernel Type: Second-Order Gaussian
No. Continuous Explanatory Vars.: 1
No. Continuous Dependent Vars.: 1

where σ_j is an adaptive measure of spread of the continuous variable defined as min{standard deviation, interquartile range$/1.349$}, n the number of observations, P the order of the kernel, and l is the number of continuous variables in the model, that is $j = 1$, $l = 2$ and $P = 2$, in our example.

The function *npcdens* returns a *condensity object*, similarly the function *npcdist* returns a *condistribution object*. We consider a grid of 100 values in the range of the covariate Length and, for each, we evaluate the estimate of the conditional hazard function on a grid of 100 values of the lifetime that we define ranging between the minimum and the maximum values. We first consider the defined bivariate grid for constructing the conditional density with the uncensored data by means of the function *npcdens*. Then, we obtain the corresponding conditional distribution function over the same grid of points with the whole dataset (censored or not) and using the function *npcdist*. We create two R objects that contain respectively the values of the estimated conditional density and conditional distribution values by calling at the function *fitted*. With the corresponding returned objects, we compute the estimated conditional hazard function at each point of the grid by dividing the estimated conditional density values multiplied by the factor $n_1/n = 0.03708$, between the one minus the estimated distribution function. The results are displayed in Fig. 7.8

The details concerning the estimators of the conditional density and conditional distribution functions are reported in Table 7.4

Example Insulating Fluid Data [46]

We consider again the data analyzed in the example of Sect. 7.2.1 to illustrate the case in which the covariate is a categorical variable. Specifically, in this case the covariate Voltage is a factor with 4 levels. We proceed in the same way as explained in the previous example to obtain an estimate of the conditional hazard

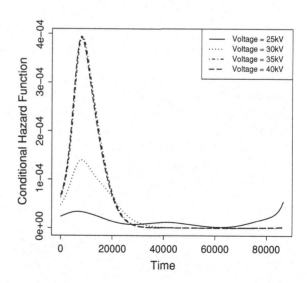

Fig. 7.9 Insulating fluid data revisited

function of failure time given the level of voltage. The results that we present in Fig. 7.9 correspond to the evaluation of the estimated conditional hazard functions for each level of voltage in a common grid of values defined into the range of the time variable. The results obtained here, where no model is assumed, could be taken into account in the discussion following the results obtained after analyzing these data in the example of Sect. 7.2.1.

Remark 7.3 Bandwidth Selection

Every nonparametric function estimation problem inherits a bandwidth (smoothing parameter) selection issue. As discussed in Chap. 1, in the case of failure rate estimation and density estimation, there are several well-established bandwidth selection methods that have been shown to possess very desirable asymptotic properties. However, in the regression models with lifetime data, bandwidth selection procedures are still in its infancy. In fact, except for a couple of ad hoc methods, there are no theoretically sound bandwidth selection criteria in the literature.

Wells [63] discusses the bandwidth selection for estimating the baseline hazard rate $\lambda_0(t)$ based on a sample $\{(Y_i, \mathbf{X}_i, \delta_i), i = 1, .., n\}$ in the Cox model. The proposed technique is based on minimizing an estimator of the local Mean Squared Error of the baseline estimator. In particular, the MSE ($M(t, b)$) of the baseline hazard rate at a point t estimated with a bandwidth b is written as

$$M(t, b) = B^2(t, b) + \text{Var}(t, b)$$

where $B(t, b)$ is the bias and $\text{Var}(t, b)$ is the variance of the estimator of $\lambda_0(t)$. The bias term $B(t, b)$ can be estimated as

$$\widehat{B}(t, b) = \int_0^1 \frac{1}{b} k\left(\frac{t-u}{b}\right) \widehat{\lambda}_{0n}(u, \hat{\beta}) \, du - \widehat{\lambda}_{0n}(u, \hat{\beta})$$

where $\widehat{\lambda}_{0n}(\cdot, \hat{\beta})$ is a pilot estimator of the baseline hazard rate using a bandwidth that satisfies the rate $O(n^{-1/5})$. Here, k is the kernel function. The term $Var(t, b)$ has been estimated using an argument given in Karr [30].

In particular, if we define $N(t) = \sum_{i=1}^{n} I[Y_i \leq t]$ and let $D_i(t) = 1$ if the ith individual is at risk at t and $D_i(t) = 0$ otherwise, an estimator of $\text{Var}(t, b)$ can be written as

$$\widehat{\text{Var}}(t, b) = \frac{1}{b^2} \int \left[J(u) K\left(\frac{t-u}{b}\right) / S^0(u, \hat{\beta}) \right]^2 \, dN(u).$$

Here,

$$S^0(t, \beta) = \frac{1}{n} \sum_{i=1}^{n} D_i(t) e^{\beta' \mathbf{X}_i}.$$

As argued by Wells [63], the estimator $\widehat{M}(t,b) = \widehat{B}^2(t,b) + \widehat{Var}(t,b)$ is consistent for $M(t,b)$, and therefore a minimization of $\widehat{M}(t,b)$ at each point t yields a locally optimal bandwidth for estimating λ_0 at t. The pilot estimator for the λ_0 can be obtained using a cross validated bandwidth.

The other available method of bandwidth selection is a plug-in method proposed by Spierdijk [56]. They consider the general regression model where the hazard rate is expressed as $\lambda(t|\mathbf{X})$ incorporating the covariate $\mathbf{X}$. They propose a very tedious yet an ad hoc method that involves an arbitrary linear model relating Ys to the covariates via some normality assumptions.

While the above methods give working data-based bandwidths, there are no established performance markers. Therefore, the use of these methods may not be superior to a general cross-validation method. As mentioned, the optimal global and local bandwidth selection issue for hazard regression remains an open problem.

References

1. Aalen OO (1980) A model for nonparametric regression analysis of counting processes. Lecture Notes in Statistics. vol 2, Springer, New York, pp 1–25
2. Aalen OO (1989) A linear regression model for the analysis of life times. Stat Med 8:907–925
3. Aalen OO (1993) Further results on the nonparametric linear regression model in survival analysis. Stat Med 12:1569–1588
4. Akaike H (1970) Statistical predictor identification. Ann Inst Stat Math 22:203–217
5. Andersen PK, Borgan Ø, Gill RD, Keiding N (1993) Statistical models based on counting processes. Springer, New York
6. Boos DD, Stefanski LA, Wu Y (2009) Fast FSR variable selection with applications to clinical trials. Biometrics 65:692–700
7. Breslow NE (1975) Analysis of survival data under the proportional hazards model. Int Stat Rev 43:45–58
8. Carrión A, Solano H, Gámiz ML, Debón (2010) A evaluation of the reliability of a water supply network from right-censored and left-truncated break data. Water Resourc Manage, doi:10.1007/s11269-010-9587-y
9. Chen YQ, Jewell NP (2001) On a general class of semiparametric hazards regression models. Biometrika 88(3):687–702
10. Chen YQ, Wang MC (2000) Analysis of accelerated hazards models. J Am Stat Assoc 95:608–618
11. Cho HJ, Hong S-M (2008) Median regression tree for analysis of censored survival data. IEEE Trans Syst Man Cybern A Syst Hum 38(3):715–726
12. Clayton D, Cuzik J (1985) The EM-algorithm for Cox's regression model using GLIM. Appl Stat 34:148–156
13. Cox DR (1972) Regression models and life-tables (with discussion). J R Stat Soc B 34:187–220
14. Crowder MJ, Kimber AC, Smith RL, Sweeting TJ (1991) Statistical analysis of reliability data. Chapman and Hall, London
15. Dabrowska DM (1997) Smoothed Cox regression. Ann Stat 25(4):1510–1540
16. Ebrahimi N (2007) Accelerated life tests: nonparametric approach. In: Ruggeri F, Kenett R, Faltin FW (eds) Encyclopedia of statistics in quality and reliability. Wiley, New York

17. Efron B (1967) Two sample problem with censored data. In: Proceedings of the fifth Berkeley symposium on mathematical statistics and probability, vol IV, pp 831–853
18. Fan J, Gibels I, King M (1997) Local likelihood and local partial likelihood in hazard regression. Ann Stat 25(4):1661–1690
19. Fan J, Li R (2001) Variable selection via nonconcave penalized likelihood and its oracle properties. J Am Stat Assoc 96:1348–1360
20. Fan J, Li R (2002) Variable selection for Cox's proportional hazard model and frailty model. Ann Stat 30:74–99
21. Fan J, Lin H, Zhou Y (2006) Local partial-likelihood estimation for lifetime data. Ann Stat 34(1):290–325
22. Fan J, Yao Q, Tong H (1996) Estimation of conditional densities and sensitivity measures in nonlinear dynamical systems. Biometrika 83(1):189–206
23. Fox J (2008) Cox proportional-hazards regression for survival data. Appendix to an R and S-PLUS companion to applied regression. Sage Publications, London
24. Gentleman R, Crowley J (1991) Local full likelihood estimation for the proportional hazards model. Biometrics 47:1283–1296
25. González-Manteiga W, Cadarso-Suárez C (1994) Asymptotic properties of a generalized KaplanMeier estimator with some applications. J Nonparametr Stat 4:65–78
26. Hall P, Racine J, Li Q (2004) Cross-validation and the estimation of conditional probability densities. J Am Stat Assoc 99:1015–1026
27. Hayfield T, Racine JS (2008) Nonparametric econometrics: the np package. J Stat Softw 27(5) http://www.jstatsoft.org/v27/i05/
28. Honoré B, Khan S, Powell JL (2002) Quantile regression under random censoring. J Econometr 109:67–105
29. Jones MC, Davies SJ, Park BU (1994) Versions of kernel-type regression estimator. J Am Stat Assoc 89:825–832
30. Karr AF (1986) Point processes and their statistical applications. Marcel Dekker, New York
31. Kalbfleisch JD, Prentice RL (2002) The Statistical analysis of failure time data, 2nd edn. John Wiley and sons, New Jersey
32. Klein M, Moeschberger W (1997) Survival analysis. Techniques for censored and truncated data. Springer, New York
33. Kleinbaum DG, Klein M (2005) Survival analysis: a self-learning text. Springer, New York
34. Koenker R (2005) Quantile regression. Cambridge University Press, Cambridge
35. Koenker R (2009) Quantreg, quantile regression http://CRAN.R-project.org/package= quantreg
36. Koenker R, Bassett GS (1978) Regression quantiles. Econometrica 46:33–50
37. Krivtsov VV, Tananko DE, Davis TP (2002) Regression approach to tire reliability analysis 78(3):267–273
38. Kvaløy JT (1999) Nonparametric estimation in Cox-models: time transformation methods versus partial likelihood methods
39. Kvaløy JT, Lindqvist BH (2003) Estimation and inference in nonparametric Cox-models: time transformation methods. Comput Stat 18:205–221
40. Kvaløy JT, Lindqvist BH (2004) The covariate order method for nonparametric exponential regression and some applications in other lifetime models. In: Nikulin MS, Balakrishnan N, Mesbah M, Limnios N (eds) Parametric and semiparametric models with applications to reliability, survival analysis and quality of life. Birkhäuser, pp 221–237
41. Lawless JF (2002) Statistical models and methods for lifetime data, 2nd edn. Wiley, London
42. Mallows CL (1973) Some comments on c_p. Technometrics 15:661–675
43. Martinussen T, Scheike TH (2006) Dynamic regression models for survival data. Springer, Berlin
44. Meeker WQ, Escobar LA (1998) Statistical methods for reliability data. Wiley, New York
45. Nelson W (1990) Accelerated testing: statistical models, test plans, and data analyses. Wiley, New York
46. Nelson W (2004) Applied life data analysis. Wiley, New Jersey

47. Nelson W (2005) A bibliography of accelerated test plans. IEEE Trans Reliabil 54(2):194–197
48. Nelson W (2005) A bibliography of accelerated test plans. Part II-References. IEEE Trans Reliabil 54(3):370–373
49. Nielsen JP (1998) Marker dependent kernel hazard estimation from local linear estimation. Scand Actuar J 2:113–124
50. Nielsen JP, Linton OB (1995) Kernel estimation in a nonparametric marker dependent hazard model. Ann Stat 23:1735–1748
51. Portnoy S (2003) Censored regression quintiles. J Am Stat Assoc 98:1001–1012
52. Royston B, Parmar MKB (2002) Flexible parametric proportional-hazards and proportional-odds models for censored survival data, with appliction to prognostic modelling and estimation of treatment effects. Stat Med 21:2175–2197
53. Scheike T, Martinussen T, Silver J (2010) Timereg: timereg package for Flexible regression models for survival data http://CRAN.R-project.org/package=timereg
54. Schwarz G (1978) Estimating the dimensions of a model. Ann Stat 6:461–464
55. Shi P, Tsai C (2002) Regression model selection—a residual likelihood approach. J R Stat Soc B 64:237–252
56. Spierdijk L (2008) Nonparametric conditional hazard rate estimation: a local linear approach. Comput Stat Data Anal 52:2419–2434
57. Therneau TM, Grambsch PM (2000) Modeling survival data. Extending the Cox model. Springer, Berlin
58. Therneau T, Lumley T (2009) survival: Survival analysis, including penalised likelihood http://CRAN.R-project.org/package=survival
59. Tibshirani R (1996) Regression shrinkage and selection via the LASSO. J R Stat Soc B 58:267–288
60. Toomet O, Henningsen A, Graves S (2009) maxLik: maximum likelihood estimation http://CRAN.R-project.org/package=maxLik
61. Wang HJ, Wang L (2009) Locally weighted censored quanitile regression. J Am Stat Assoc 104:1117–1128
62. Wang JL (2003) Smoothing hazard rates. Encyclopedia of biostatistics
63. Wells MT (1994) Nonparametric kernel estimation in counting processes with explanatory variables. Biometrika 81:759–801
64. Wu Y, Boos DD, Stefanski LA (2007) Controlling variable selection by the addition of pseudovariables. J Am Stat Assoc 102:235–243
65. Yang S (1999) Censored median regression using weighted empirical survival and hazard functions. J Am Stat Assoc 94(445):137–145
66. Ying Z, Jung SH, Wei LJ (1995) Survival analysis with median regression models. J Am Stat Assoc 90(429):178–184
67. Zhang H, Lu W (2007) Adaptive-LASSO for Cox's proportional hazards model Biometrika 94:691–703
68. Zhao Y, Chen F (2008) Empirical likelihood inference for censored median regression model via nonparametric kernel estimation. J Multivar Anal 99:215–231
69. Zou H (2006) The adaptive LASSO and its Oracle properties. J Am Stat Assoc 101:1418–1429

Index